Muhammad Musaddique Ali Rafique

**Bulk Metallic Glasses and Their Composites**

## Also of Interest

*Advanced Aerospace Materials.*
*Aluminum-Based and Composite Structures*
Haim Abramovich, 2019
ISBN 978-3-11-053756-7, e-ISBN (PDF) 978-3-11-053757-4,
e-ISBN (EPUB) 978-3-11-053763-5

*Advanced Composites Volume 1:*
*Nanocomposites.*
*Materials, Manufacturing and Engineering*
J. Paulo Davim, Constantinos A. Charitidis (Eds.), 2013
ISBN 978-3-11-026644-3, e-ISBN (PDF) 978-3-11-026742-6

*Advanced Composites Volume 2:*
*Biomedical Composites.*
*Materials, Manufacturing and Engineering*
J. Paulo Davim (Ed.), 2014
ISBN 978-3-11-026669-6, e-ISBN (PDF) 978-3-11-026748-8

*Advanced Composites Volume 5:*
*Ceramic Matrix Composites.*
*Materials, Manufacturing and Engineering*
J. Paulo Davim (Ed.), 2016
ISBN 978-3-11-035284-9, e-ISBN (PDF) 978-3-11-035300-6,
e-ISBN (EPUB) 978-3-11-038346-1

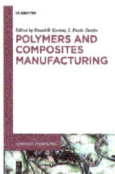

*Advanced Composites Volume 11:*
*Polymers and Composites Manufacturing.*
Kaushik Kumar, J. Paulo Davim (Eds.), 2020
ISBN 978-3-11-065193-5, e-ISBN (PDF) 978-3-11-065504-9,
e-ISBN (EPUB) 978-3-11-065212-3

*Advanced Composites Volume 12:*
*Glass Fibre-Reinforced Polymer Composites.*
*Materials, Manufacturing and Engineering*
Jalumedi Babu, J. Paulo Davim (Eds.), 2020
ISBN 978-3-11-060828-1, e-ISBN (PDF) 978-3-11-061014-7,
e-ISBN (EPUB) 978-3-11-060858-8

Muhammad Musaddique Ali Rafique

# Bulk Metallic Glasses and Their Composites

Additive Manufacturing and Modeling and Simulation

2nd Edition

**DE GRUYTER**

**Author**
Dr. Muhammad Musaddique Ali Rafique
Eastern Engineering Solutions
77 Massachusetts Avenue
Cambridge 02139
USA
ali.rafique@easternengineeringcambridge.onmicrosoft.com

ISBN 978-3-11-074721-8
e-ISBN (PDF) 978-3-11-074723-2
e-ISBN (EPUB) 978-3-11-074741-6

**Library of Congress Control Number: 2021942893**

**Bibliographic information published by the Deutsche Nationalbibliothek**
The Deutsche Nationalbibliothek lists this publication in the Deutsche Nationalbibliografie;
detailed bibliographic data are available on the Internet at http://dnb.dnb.de.

© 2021 Walter de Gruyter GmbH, Berlin/Boston
Cover image: R_Type/iStock/Getty Images Plus
Typesetting: Integra Software Services Pvt. Ltd.
Printing and binding: CPI books GmbH, Leck

www.degruyter.com

---

The author would like to dedicate this work to Abū ʿAlī al-Ḥasan ibn al-Ḥasan ibn al-Haytham (c 965–c 1040).

(father of modern optics)

# Contents

# Section 1
# Research significance and background

Discovered in 1960 by Duwartz et al. [1] at Caltech, metallic glasses emerged as a completely new class of materials exhibiting very high tensile strength, hardness, elastic strain limit and yield strength at relatively lower density as compared to steel and other high-strength alloys [2–4]. Yet, their use has not been able to get broad acceptance as a competing engineering material because of the lack of ductility and inherent brittleness of glassy structure [3]. This property becomes even more prominent at large length scales (bulk metallic glasses (BMG) – metallic glasses typically having thickness >1 mm) [5–8] as prominent catastrophic failure mechanisms (shear band) dominate [9–11]. This severely limits their application toward use in making large-scale machinery components. This disadvantage can be overpowered by inducing plasticity in a glassy structure while retaining its high strength at the same time [12–15]. This can be done by various mechanisms, including exploitation of intrinsic ability of glass to exhibit plasticity at very small (nano) length scale [16, 17], introduction of external impulse (obstacles) to shear band formation and propagation (*ex situ* composites) [18, 19], self or externally assisted multiplication of shear bands [11, 20], formation of ductile phases in brittle glassy matrix during solidification (in situ composites) [21–24] and transformation inside a ductile crystalline phase, for example, B2–B19′ transformation in Zr-based systems (stress/transformation-induced plasticity) [25–28]. The later approach (formation of ductile phase in brittle glass) takes into account the nucleation of secondary (ductile) phase either during solidification in situ [29–35] or heat treatment of solidified glassy melt (devitrification/relaxation) [36–44] and forms the basis of ductile BMG composites.

Although considerable progress has been made in advancing "as-cast" sizes in BMG and their composites, still maximum possible diameter and length, which has been produced by conventional means till date [45], is far from the limit of satisfaction to be used in any structural engineering application. This is primarily associated with mechanical cooling rate achievable as a result of quenching effect from water-cooled walls of Cu container which in itself is not enough to overcome critical cooling rate ($R_c$) of an alloy (~ 0.067 K/s [45]) to produce a uniform large bulk glassy structure. In addition to this, occurrence of this bulk glassy structure is limited to compositions with excellent inherent glass-forming ability (GFA) [46, 47]. This is not observed in compositions that are strong candidates to be exploited for making large-scale industrial structural components [26, 48–56] with higher critical cooling rates ($R_c$) (10 K/s [49]). This poses a limitation to this conventional technique and urges the need of advanced manufacturing method that does not

https://doi.org/10.1515/9783110747232-001

encompass these shortcomings. Additive manufacturing (AM) has emerged as a potential technique [57, 58] to fulfill this gap and produce BMG matrix composites [59, 60] in a single step across a range of compositions virtually covering all spectrums [61–64]. It achieves this by exploiting very high cooling rates available in a very short period transient liquid melt pool [65–67] in a small region where high-energy source laser (LSM (laser surface melting)/laser surface forming LSF (solid), selective laser melting/LENS® (powder) or electron beam melting) strikes the sample. This, when coupled with superior GFA of BMG matrix composites (BMGMC), efficiently overcomes the dimensional limitation as virtually any part carrying glassy structure can be fabricated. In addition, incipient pool formation [67] and its rapid cooling result in extremely versatile and beneficial properties in final fabricated part such as high strength, hardness, toughness, controlled microstructure, dimensional accuracy, consolidation and integrity. The mechanism underlying this is layer-by-layer (LBL) formation, which ensures glass formation in each layer during solidification before proceeding to the next layer. That is how a large monolithic glassy structure can be produced. This LBL formation also helps in development of secondary phases in a multicomponent alloy [68–70] as a layer preceding fusion layer (which is solidified) undergoes another heating cycle (heat treatment) below melting temperature ($T_m$) somewhat in the nose region of TTT (time–temperature transformation) diagram [59], which not only assists in phase transformation [41, 43] but also helps in the increase in toughness, homogenization and compaction of part. This is a new, promising and growing technique of rapidly forming metal [71], plastic [72], ceramic or composite [73] parts by fabricating a near-net shape out of raw materials either by powder method or wire method (classified on the basis of additives used). The movement of energy source (laser or electron beam) is dictated by a CAD geometry which is fed to a computer at the back end and maneuvered by computerized numerical control (CNC) [74, 75] system. Process has a wide range of applicability across various industrial sectors ranging from welding [76–81], repair [82, 83] and cladding [84–90] to full-scale part development.

However, there is dearth of knowledge about exact mechanisms of formation (nucleation and growth (NG) and/or liquid–liquid transition [91–93]) of ductile-phase dendrites or spheroids in situ during solidification of BMGMC happening inside liquid melt pool of AM which is essential to further advance improvement in the process and suggest its optimization. Modeling and simulation techniques especially those employing finite element methods (phase field [94–99], cellular automaton finite element (CAFE) [100–105] and their variants) at part scale are very helpful in explaining the evolution of microstructure and grain size development in metals and alloys. They have been extensively used in predicting solidification behavior of various types of melts during conventional production methods [105–108]. However, their use in

AM applications [109–112] especially related to BMGMC is still in its infancy. Virtually no effort has been made to understand nucleation and growth of ductile crystalline phase dendrites or spheroids in situ during solidification in BMGMC by modeling and simulation. A step forward is taken in this study to address these gaps and bring together the strengths of different techniques and methodologies at one platform. An effort is made to form ductile BMGMC by considering the following advantages:

a. Material chemistry: This is a multicomponent alloy. Its GFA is used as a measure to manipulate composition and vice versa.

b. Solidification process: Liquid melt pool formation – its size, shape and geometry, and role of number density, size and distribution of ductile phase result in glassy alloy matrix. It is taken as a function of type, size and amount of nucleates (inoculant).

c. AM: Use of very high cooling rate inherently available in the process to (a) not only form glassy matrix but use liquid melt pool formed at very high temperature to trigger nucleation (liquid–solid transformation) of ductile phase in the form of dendrites or spheroids from within the pool "in situ" (this is done by controlling machine parameters in such a way that optimized the cooling rate satisfying narrow window of "quenching" BMGs is achieved) and (b) take advantage of heating (heat treatment) of preceding layer to trigger solid–solid transformation (devitrification) again to form ductile phase and achieve homogeneity, consolidation and part integrity eliminating the need of post processing or after treatment.

d. Modeling and simulation: Strong and powerful mathematical modeling techniques based on
   a. transient heat transfer for "liquid melt pool formation as a result of laser–matter interaction" and
   b. its "evolution–solidification" by
      i. deterministic (modified classical nucleation theory, Johnson–Mehl–Avrami–Kolmogorov correction and Rappaz modification) or
      ii. stochastic/probabilistic (3D CAFE model for nucleation and growth (solute diffusion and capillary action driven))
      modeling of microstructure evolution, and grain size determination of ductile phase equiaxed dendrites or spheroids in glassy melt will be used to simulate the conditions in liquid melt pool of BMGMC during AM. Effect of number density, size and distribution of ductile phase dendrites or spheroids will be evaluated/verified using simulation of melt pools developed using different value of aforementioned parameters.

**Note:**

a.  AM methods can also be classified on the basis of energy source used (i.e., laser based or electron beam based).

b.  Heating in the nose region of TTT diagram can also trigger assimilation of free volume (relaxation), whose effect is not taken into account in this study due to difference in the mechanism in which it occurs. It does not constitute any chemical reaction.

# Section 2
# Bulk metallic glasses (BMGs) and bulk metallic glass matrix composites (BMGMCs)

## 2.1 Metallic glasses (MG) and bulk metallic glasses (BMGs)/monoliths

Metallic glasses (MG) [5] may be defined as "disordered atomic-scale structural arrangement of atoms formed as a result of rapid cooling of complex alloy systems directly from their melt state to below room temperature with a large undercooling and a suppressed kinetics in such a way that the supercooled state is retained/frozen" [113–116]. This results in the formation of a "glassy structure." The process is very much similar to inorganic/oxide glass formation in which large oxide molecules (silicates/borides/aluminates/sulfides and sulfates) form a regular network retained in its frozen/supercooled liquid state [117], the only difference being MG are comprised of metallic atoms rather than inorganic metallic compounds. Their atomic arrangement is governed by mismatch of atomic size and quantity (minimally three) [118] (described in the next section) and is based on short-range order (SRO) [119–121] to medium-range order (MRO) [122–124] or long-range disorder [4] (unlike metals – well-defined long-range order) and can be explained by other advanced theories/mechanisms (frustration [125], order in disorder [123, 125, 126] and confusion [127]). Important features characterizing them are their amorphous microstructure and unique mechanical properties. Owing to absence of dislocations, no plasticity is exhibited by BMGs. This results in very high yield strength and elastic strain limits as there is no plane for material to flow (by conventional deformation mechanisms). From a fundamental definition point of view, MG are typically different from bulk metallic glass (BMG) in that the former has fully glassy (monolithic) structure for thicknesses less than 1 mm, while the later is glassy (monolithic) is greater than 1 mm [6, 7]. To date the largest BMG made in "as-cast" condition is 80 mm diameter and 85 mm in length [45]. There are reports of making large thin castings as casing of smart phones, but they are typically less than 1 mm [10]. Furthermore, they are characterized by special properties such as glass-forming ability (GFA) and metastability (which will be described in the subsequent sections).

## 2.2 Three laws

The formation and stability of BMG (even in metastable condition) is described by their ability to retain glassy state at room temperature. Although the understanding of glass and glassy structure was established much earlier, it was very difficult to

https://doi.org/10.1515/9783110747232-002

form homogeneous, uniform glassy structure across whole section thickness at room temperature until recently. Only alloys of very narrow compositional window cooled at extremely high cooling rate can form glassy structure [1, 5, 6, 128, 129]. Any deviation from any of these parameters severely hampers the retention of glassy state and crystallization occurs [130–132]. This property is known as GFA [133]. This is the single most important property in MG family of alloys which governs their formation and evolution. GFA has been increasingly studied, and considerable progress has been made in its improvement [134–137] by alterations in both composition and window of processing condition [4, 138, 139]. Now, alloys having multicomponent composition can be cast in glassy state even at slow cooling rate owing to their superior GFA [49, 135, 140–144], which in turn is governed by various theories [137, 140, 145–156] and analytical models [157, 158].

Fundamentally, research over the period of time has yielded three basic laws which are now considered universal for forming any BMG system [118]. These are described below. Any glass-forming system consists of elements which must
1.  Be three in number (at minimum). (Elements >3 are considered beneficial.)
2.  Differ in their atomic size by 12% among three elements. (Atoms of elements with large size are considered to exhibit superior GFA.)
3.  Have negative heat of mixing among three elements. (This ensures tendency to demix or confuse [127] ensuring retention of glassy structure at room temperature.)

This results in a new structure with a high degree of densely packed atomic configurations, which, in turn, results in completely new atomic configuration at a local level with long-range homogeneity and attractive interaction. In general, BMG or BGA are typically designed around alloy systems that exhibit (1) a deep eutectic, which decreases the amount of undercooling needed to vitrify the liquid, and (2) alloys that exhibit a large atomic size mismatch, which creates lattice stresses that frustrate crystallization [118]. An important way to arrive at optimum glass-forming composition and then selecting alloying elements is based on proper choice of eutectic or off-eutectic composition, diameter and heat of mixing [4]. These laws were first proposed by Douglas C. Hoffmann and his supervisor Abdulah S. Jabri at Caltech [118] and Prof. Akisha Inoue at WPI – IMR, Tohoku University, Japan [4], independently.

## 2.3 Classification – various approaches

BMG may be classified into various categories based on their constituents (i.e., elements), chemical composition and makeup, type, atomic size difference, heat of mixing and period of constituent elements. These have been elaborately discussed in various studies [631–632], which are enumerated below.

### 2.3.1 Based on elements and their types

As proposed by Professor Inoue [4, 159, 160], BMG can be broadly **classified** into **three** categories (Figure 1):

1. metal–metal type
2. Pd–metal–metalloid type
3. metal–metalloid type

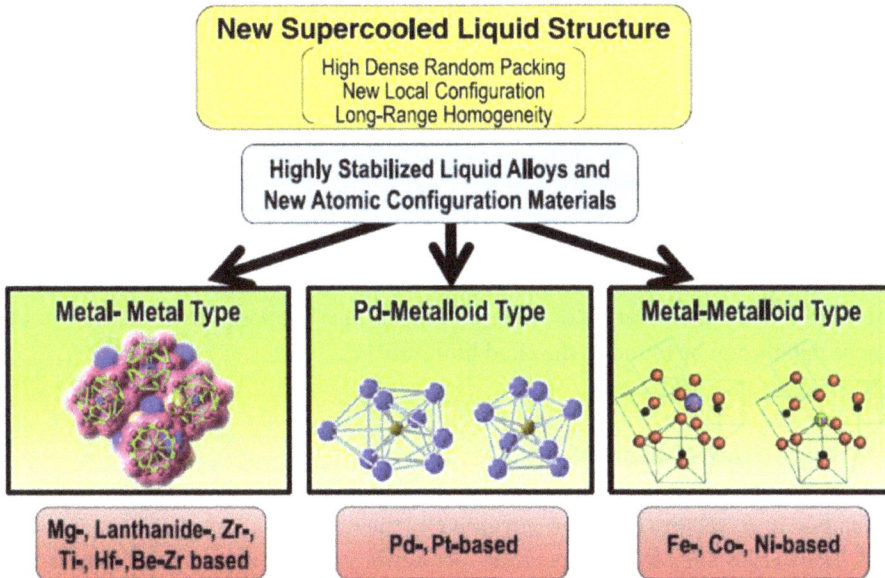

**Figure 1:** Classification of bulk glassy alloys [4, 159].

This classification is based on ease with which one group of metals react with other group to finally evolve as a glassy structure, which in turn is chosen by various rules such as chemical affinity, atomic size and electronic configuration. Their proposed atomic arrangement, size and crystal structure is shown in Figure 1. *Metal–metal-type* glassy alloys are composed of icosahedral-like ordered atomic configurations. They are exemplified by Zr-Cu-Al-Ni and Zr-Cu-Ti-Ni-Be-type systems. *Pd–transition metal–metalloid*-type glassy alloys consist of high dense packed configurations of two types of polyhedra of Pd-Cu-P and Pd-Ni-P atomic pairs. Their typical examples are Pd-Cu-Ni-P systems. *Metal–metalloid–type* glassy alloys have network like atomic configurations in which a disordered trigonal prism and an anti-Archimedean prism of Fe and B are connected with each other in face and edge shared configuration modes through glue atoms of Ln and ETM of Zr, Hf and Nb. Their typical examples are Fe-Ln-B and Fe-(Zr, Hf, Nb)-B ternary systems. These icosahedral-, polyhedral- and network-like ordered atomic configurations can effectively suppress the long-range rearrangements of the constituent elements, which are necessary for the onset of crystallization process. Among the three structures described, second and third types have similarity that they both contain trigonal prism structure but are different in that later forms a well-developed connected structure of prisms by sharing their vertices and edges which results in a highly stabilized supercooled liquid leading to the formation of BGA even at very slow cooling solidification processes [4]. From **engineering** standpoint, bulk glassy alloy (BGA) adopts another system of classification which is based on their applicability. They are classified into seven types which in turn are grouped into two main types based on their behavior in phase diagrams. These are described as follows:

a.  host metal base type: Zr-Cu-Al-Ni, Fe-Cr-metalloid, Fe-Nb-metalloid and Fe-Ni-Cr-Mo-metalloid systems and
b.  pseudo-host metal base type: Zr-Cu-Ti-Ni-Be, Zr-Cu-Ti-(Nb, Pd)-Sn, and Cu-Zr-Al-Ag systems

It can be observed that Fe and Zr comprise most important materials for practical use. Further subclassification of Zr-based BMG is also proposed by Professor Inoue whose details can be found in the cited literature [4].

## 2.4 Important characteristics

Formation and stability of BMG is governed by their ability to form complex network and then retain this at a temperature below room temperature. This is best described by its intrinsic properties specific to these alloy systems. These are GFA and metastability.

### 2.4.1 Glass-forming ability

As described in Section 2.2, GFA may be defined as "inherent, intrinsic ability of a multicomponent system to consolidate in state of low energy in such a way that glass formation is promoted and crystallisation is retarded." This single unique parameter is effectively used to identify and design a range of glassy alloys. The GFA of a melt is evaluated in terms of the critical cooling rate ($R_c$) for glass formation, which is *the minimum cooling rate necessary to keep a constant volume of melt amorphous without precipitation of any crystals during solidification* [161–165]. In addition to this, they must possess inherent resistance against crystallization; that is, their atomic configuration should be such that they should not favor its rearrangement in regular crystallographic patterns. GFA is a strong function of another parameter known as "overall cooling power" or "strength of quench." Generally

$$\text{GFA} \propto \text{strength of quench} \tag{1}$$

which means the higher the quenching power, better will be the ability of a material to form glass. However, this is not a hard-and-fast rule and exceptions exist [49, 135, 140–144] (as described in Section 2.2). For example, in a well-defined multi-component system, for example, Zr-Ti-Cu-Ni-Be [49], BMG can be formed even at slower cooling rate. This is because, above-mentioned two criterions are effectively fulfilled in these BMG's while in others, for example, Ti- and Cu-based BMG, glassy structure can only form in relatively thin sections (because of very high cooling rates experienced there) – which is essential for glass formation in these systems. Also, these systems do not exactly meet above criterion and deviations exist which promote their inability to form glassy structure even upon fast cooling. Metals which most commonly account for formation of BMG are early transition metals (ETM) and late transition metals (LTM) [159, 160, 166]. From phase development point of view, they always include a eutectic point with the lowest liquidus temperature. An important factor to design these alloys is to choose a composition exhibiting a lower liquidus temperature in the vicinity of eutectic point. Although variants exist (off-eutectic compositions) [167–171], this method is effective to an appreciable extent for the design of BGA/BMGs [4].

There have been different theories the way GFA has been predicted over years. For example, David Turnbull in his classical paper [129] mentioned the use of reduced glass transition temperature ($T_{rg}$) where it is defined as the ratio of glass transition temperature ($T_g$) and liquidus temperature ($T_l$)

$$T_{rg} = \frac{T_g}{T_l} \tag{2}$$

This still has been the basic method of determining GFA to a large extent. However, there have been limitations around it and there were other theories which were predicted. For example, the use of *supercooled liquid region* $\Delta T_x$ (the temperature difference between the onset crystallization temperature $T_x$ and glass transition temperature $T_g$) [166].

$$\Delta T_x = T_x - T_g \tag{3}$$

The $y$ parameter [145] is defined as

$$y = \frac{T_x}{(T_g + T_l)} \tag{4}$$

None of these alone or combined is good enough to predict the GFA of BMGs [133, 134, 137, 172] and the GFA remain a function of alloy composition to a large extent empirically which keeps changing [122, 136, 141, 142, 149, 151, 173]. Following diagrams can be effectively used to arrive at nearest possible composition at which BMG alloy formation is expected in the mentioned ternary (Figure 2) and quaternary systems (Figure 3).

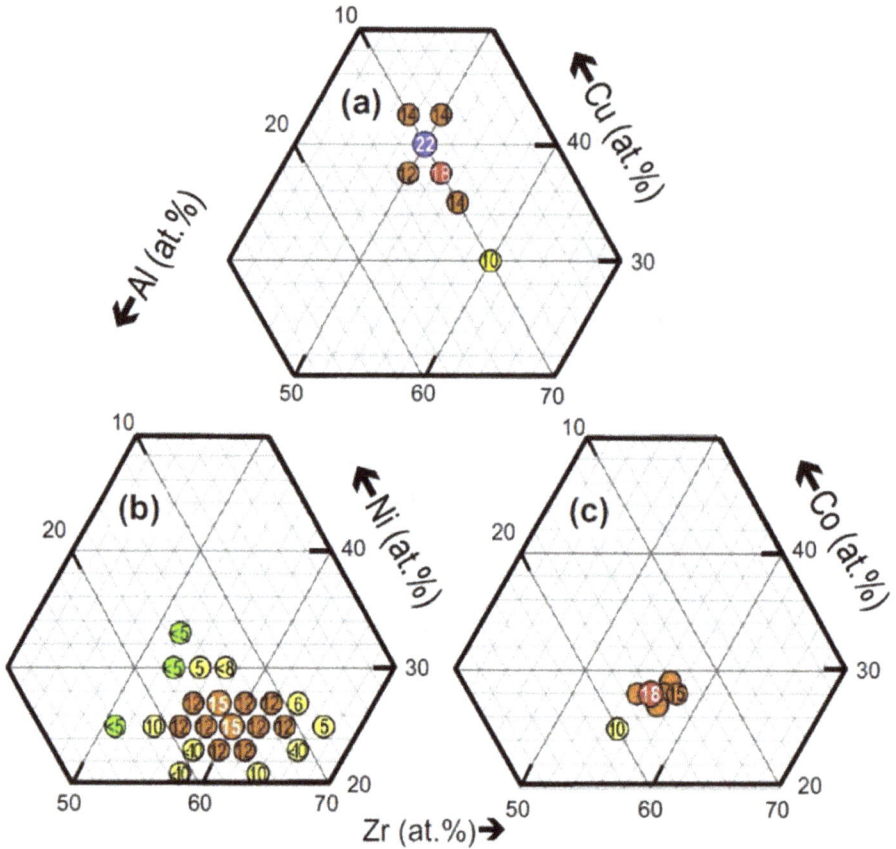

**Figure 2:** Composition range in which the BGA are formed by the copper mold casting method and the composition range of the maximum diameter of cast glassy alloy rods in Zr–Al–Cu, Zr–Al–Ni and Zr–Al–Co systems [4].

From a phase transformation point of view, they follow ternary phase diagrams more predominantly than binary diagrams because of constraint posed by necessity of having three elements. Their mechanical properties can also be explained on the basis of ternary phase diagrams more effectively. When used in conjunction with above compositional contrast diagrams (Fig 2 and 3), these can effectively predict a suitable alloy system which will show superior GFA along with a set of mechanical properties [4].

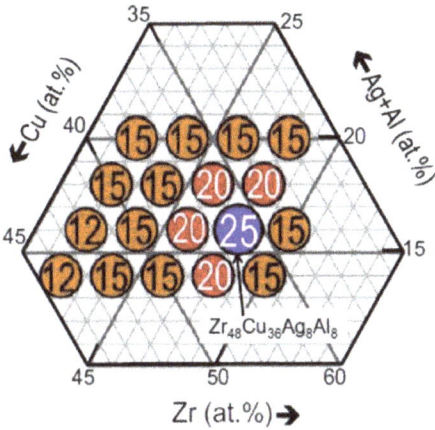

**Figure 3:** Compositional dependence of maximum diameter of Zr–Cu–Al–Ag glassy alloys produced by copper mold casting [4].

### 2.4.2 Metastability

Another important characteristic of these glass forming systems is their composition which also describes their inhomogeneity and metastability. They are not cooled to room temperature following equilibrium phase diagram, but their formation and evolution is governed by nonequilibrium diagrams also known as time–temperature transformation (TTT) diagrams [174]. This gives rise to metastable structures resulting from very high cooling rates [175–177]. These metastable structures are reasons of extremely high strength of these systems. Upon heating, BMG relax their structural disorder/misfit and give rise to ordered structures. This process is known as devitrification [178–183]. This also helps further to explain and understand the development and formation of quasicrystals (QC) [154, 184–188] and ductile phases (e.g., B2 CuZr, β-Zr) which are responsible for increase in ductility and toughness of BMG and their composites (BMGMC). It is also very important in defining the behavior of BMGMC in additive manufacturing (AM) as material undergoes repeated thermal cycles which vitrify the liquid and devitrify the glass.

## 2.5 Limitations

Despite their advantages and extremely high strength, MG and their bulk counterparts suffer from following limitations:

a. They have very poor ductility [3, 189–191]. They do not exhibit any plasticity under tension and exhibit little plastic behavior under compression [192–194].

b. They have very poor fracture toughness [13, 195–201]. This severely limits their engineering applications as they cannot absorb effects of load or cannot transfer stresses safely and fail in a catastrophic manner [202].

Progress has been made during recent years to overcome these problems, but still experimental results and values obtained so far are not of considerable practical significance and have poor reproducibility which render them unsatisfactory for any practical use [203–205].

## 2.6 Ductile BMG

Owing to difficulties encountered during the use of "as-cast" BMG especially for structural applications, schemes were devised from the very early days of BMG research for the increase of ductility in these alloys. In the beginning, efforts were made to increase the plasticity by dispersing controlled porosity [206] but these efforts did not proceed far because of nonpractical nature of method and other unwanted problems developed in the structure. Then, the focus was directed to address this problem by basic mechanisms of plasticity and plastic deformation. For example, if progression of shear band could be hindered (just like dislocation motion hindrance in crystalline alloys) by impeding its motion, a substantial increase in ductility could be achieved. This is achieved by two fundamental mechanisms: (a) increased number of shear bands increases the obstacles ("arrests") to the paths of material flow. Hence, it would be difficult for the material to flow [207–214]. (b) Strain energy dissipation resulting from shear band formation at the interface between crystalline phase and amorphous matrix. This gives rise to new processes of shaping/forming by controlled application of force (thermoplastic forming, TPF) [215, 216]. This new method of fabricating BMG is also known as superplastic forming [217] which was tried as far as 10 years ago. Further techniques consisted of (1) ex situ introduction of second-phase reinforcements (particles [19, 218, 219], flakes [220], fibers [221–223], ribbons [224], whiskers [225, 226], etc.) which offer a barrier to movement of shear bands along one plane and provide a pivot for their multiplication, (2) In situ nucleation and growth (NG) of second-phase reinforcements in the form of equiaxed dendrites which are ductile in nature thus not only provide means of increase in ductility by themselves but offer a pivot for multiplication of shear bands (explained in the next section) [227, 228] (3), reducing the size of glass to nanometer and ductile phase to micrometer [27], (4) making the plastic front (local plastically deformed region ahead and around shear band) of shear bands to match with difficult plane of flow in crystal structure of ductile phase, thus creating easy path for shear band to multiply (not yet investigated idea of author) and (5) heating the alloy to cause temperature induced structural relaxation (devitrification) [178, 180–182, 229]. The drive for all these mechanisms is different. For example, it is natural that shear bands are responsible for the catastrophic failure of BMGs [230] and any hindrance to their motion or simply increase in their number all spread across the volume of material would cause a difficulty with which they move (along one direction at very high speed) causing abrupt failure. This gives rise to fundamental mechanisms of toughening [13, 231]. Similar effect could be achieved by external addition to (ex situ), or internal manipulation of (in situ) structure of material. Of these, only structural relaxation was first envisaged as dominant mechanism for increase of ductility. It was known thermodynamically and observed numerically [232] and experimentally [233–236] since early days that structurally constrained glass relaxes during heating known as "devitrification." The drive for devitrification [178, 179] did not come out of an ingenious effort but was a natural impulse as BMG possess natural tendency to undergo release of stresses with formation of new structures (phases) (solid state phase transformations) when subjected to temperature effect similar to heat treatment for crystalline metallic alloys. This resulted in a new class of BMG called **ductile BMG** [237–245]. The research on other mechanisms was adopted, or is envisaged to be investigated, with passage of time, giving rise to more versatile materials known as ductile BMG composites (explained in the following).

## 2.7 Ductile bulk metallic glass matrix composites

As minutely introduced in the previous section, a significant improvement in mechanical properties of BMG was reported for the first time in 2000 by Professor William's Group at Caltech [12] when they successfully incorporated the ductile second-phase reinforcements in glassy matrix in the form of precipitates in situ formed during solidification, thus giving birth to so-called family of in situ dendrite/MG matrix composites. These materials are formed as a result of conventional solute partitioning mechanisms as observed in other metallurgical alloys resulting in a copious formation of ductile phase (Ti-Zr-Nb β in case of Ti-based composites [12], B2 Cu-Zr in case of Zr-based composites [246–251] or transformed B2 (B19′ martensite) in case of Zr-Cu-Al-Co shape memory BMGMC (a special class of BMGMC) [21, 249, 252–256]) predominantly (not always) in the form of three–dimensional (3D) dendrites or spheroids emerging directly from liquid during solidification. Devitrification in these alloys can be explained by "phase separation" before solid state transformation or "quenched in" nuclei [257–261]. This is another very important route for the fabrication of these alloys. They also comprise of a family of BMG composites which are formed by more advanced transformations mechanisms (liquid state phase separation) [262–265] which has recently become observable owing to more advanced characterization techniques using synchrotron radiation [266–269] and container less levitated sample solidification in micro and zero gravity conditions [92, 270]). These render them special properties (enhanced plasticity and compressive strength) not otherwise attainable by other conventional processing routes or in simple binary and ternary compositions. This, however, is seldom the case and is not readily observed as compared to solid state phase separation [262] which is dominant mechanism in these alloys. More advanced mechanisms of forming these materials are by local microstructural evolution by phase separation right at shear bands [135]. It narrates that solid–solid phase separation (spinodal decomposition) occurs at the onset of shear band which is the cause of microstructural evolution.

Few notable class of alloys in these types of ductile composites are Ti-based BMGMCs [55, 56, 271–276], Ti-based shape memory BMGMC [277], Zr-Cu-Al-Ti [278, 279], Zr-Cu-Al-Ni [52] and Zr-Cu-Al-Co shape memory BMG composites [51]. Each has its own mechanisms of formation and individual phases are formed by liquid–solid (L–S) or solid–solid (S–S) phase transformations.

## 2.8 Production methods – mechanisms perspective

From mechanisms perspective, production and processing of BMG matrix composites can be characterized into two fundamental types;

### 2.8.1 Liquid–solid transformation (solidification)

This mechanism is typically characterized by solidification of melt from liquid state to room temperature. The resulting metal matrix composite may consist of melt which carry solid particles suspended, or homogeneously mixed in the melt prior to, or during casting as a result of very chemical nature of alloy (limit of solid solubility enabling formation of substitutional or interstitial solid solution upon solidification). These are called ex situ and in situ BMGMC, respectively.

#### *a.* Ex situ formation

These are methods of production in which external particles (reinforcements) in various forms (particles, flakes, fibers and whiskers (Section 2.6)) are introduced in the bulk melt and are homogenized by various means before final shape casting. These are versatile in a sense that properties could be manipulated by controlling type, size, shape and amount of external reinforcements. Furthermore, various types of numerical and/or statistical means (permutations and combinations) could be applied to control the material processing and fabrication. Melt infiltration is one of popular routes for the fabrication of these composites. This is one of well-established techniques in ceramics processing [280]. This was one of first method adopted again at Caltech [218] in 1997–98 to fabricate their world famous Vitreloy$^{TM}$1 ($Zr_{41.2}Ti_{13.8}Cu_{12.5}Ni_{10.0}Be_{22.5}$) as matrix with continuous ceramic (SiC and carbon), and metal (tungsten (254 µm Ø), carbon steel (AISI 1080) (254 µm Ø) (in most of the cases), stainless steel, molybdenum, tantalum, nickel, copper, and titanium) wires cut into 5 cm length as well as with loose tungsten (W) powders and sintered silicon carbide (SiC) particulate preforms as reinforcements. They infiltrated different specimens with aforementioned reinforcements and casted rods in size 5 cm in length and 7 mm in diameter. The process was carried out in vacuum induction furnace with titanium gettered Ar atmosphere. The starting metals were high-purity (≥ 99.5% metal basis) research-grade materials. The composite fabricated were quenched after infiltration. A schematic of setup used for this purpose is shown in Figure 4.

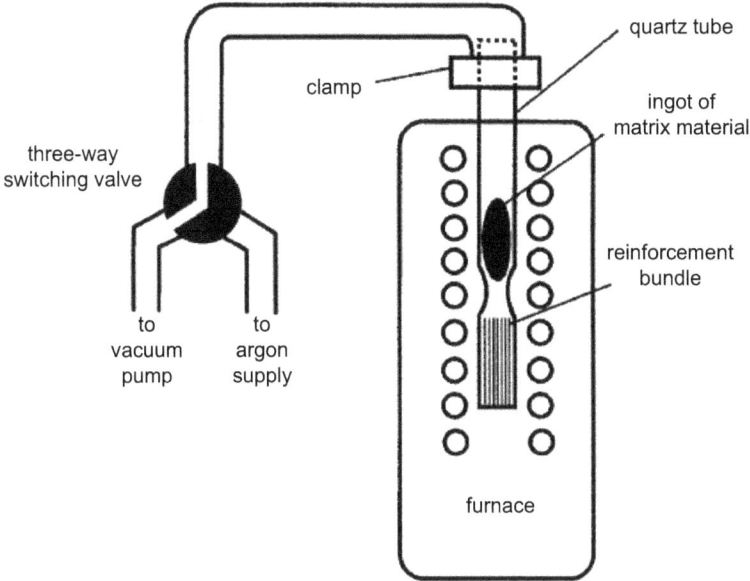

**Figure 4:** Schematic of setup for melt infiltration [218].

The resulting composite was analyzed with X-ray diffraction (XRD) and scanning electron microscopy (SEM). The measured porosity was ≤3% and matrix was about 97% amorphous material. The details of procedure could be found in Reference [218]. They also carried out the mechanical (compressive and tensile) and structural (SEM) characterization and fractography of the composite developed in another study [221]. It was found that tungsten reinforcement increased compressive strain to failure by over 900% compared to the unreinforced monolithic Vitreloy$^{TM}$1. A definite increasing near linear relationship was observed between fiber volume fraction ($V_f$) and elastic modulus ($E$) (Figure 5a) and yield stress (Figure 5b). The increase in compressive toughness was attributed to fibers restricting propagation of shear bands which in turn promotes generation of further multiple shear bands and additional fracture surface area. There is direct evidence of viscous flow of the MG matrix inside regions of the shear bands. Samples reinforced with steel were found to have an increased tensile strain to failure and energy to break by 13% and 18%, respectively. Reason for the increase in tensile toughness was found to be ductile fiber delamination, fracture and fiber pullout as observed by fractography under SEM (Figure 6).

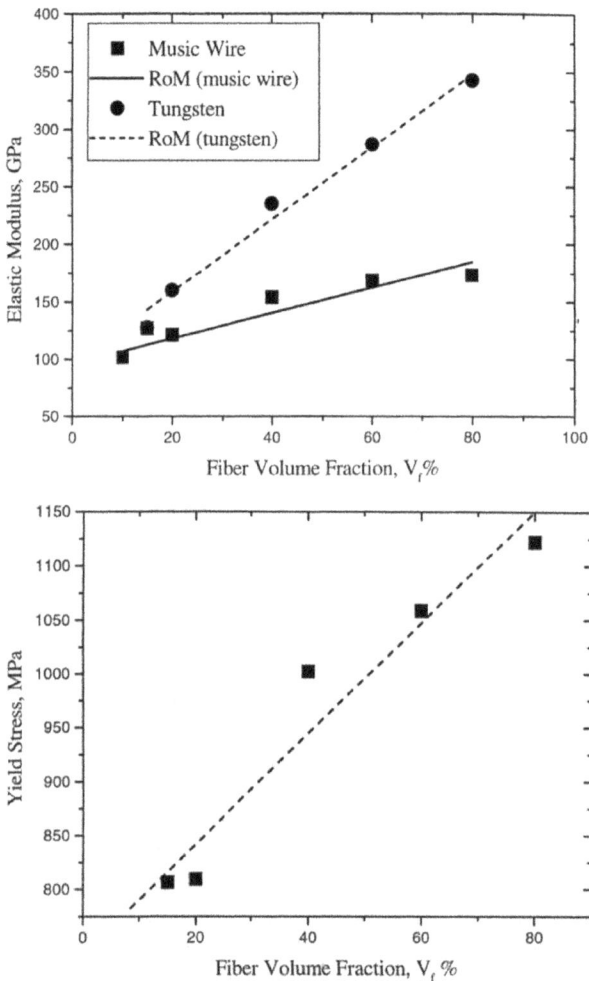

**Figure 5:** (a) Relationship between volume fraction ($V_f$) of fiber (percentage) and elastic modulus expressed in GPa. Dashed lines are rule of mixture (RoM) modulus calculations [221].
(b) Relationship between volume fraction ($V_f$) of fiber (percentage) and yield stress (MPa) [221].

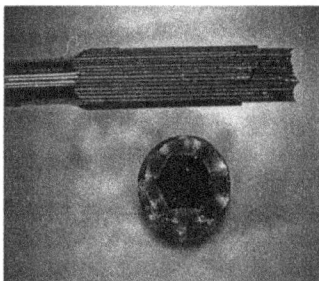

**Figure 6:** Fiber pulloutof 80%steel wire/Vitreloy[TM]1 composite tensile specimen [221].

In a further study [281], they modified the alloy composition to $Zr_{57}Nb_5Al_{10}Cu_{15.4}Ni_{12.6}$ and repeated the same procedure of melt infiltration without AISI 1080 carbon steel wire but with Ta particles. Other reinforcements (SiC and W) remained same in varying volume fractions. The samples were tested under compression and tension as done previously. It was found that compressive strain to failure increased by more than 300% compared with the unreinforced $Zr_{57}Nb_5Al_{10}Cu_{15.4}Ni_{12.6}$ monolithic BMG and energy to break of the tensile samples increased by more than 50%. A typical fracture surface of $W/Zr_{57}Nb_5Al_{10}Cu_{15.4}Ni_{12.6}$ specimen is shown in Figure 7, indicating direction of shear band propagation.

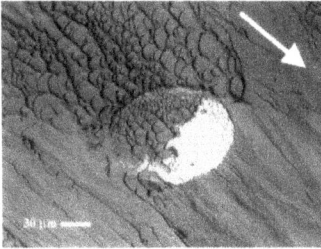

**Figure 7:** Fracture surface of W/Vitreloy™1 showing direction of shear band propagation [281].

This study was extended in 2001 [282] by another postdoctorate Ms. Haein Choi-Yim and she performed quasistatic as well as dynamic tests. The aim was to manufacture so-called kinetic energy penetrators (KEP) under a program funded by US Army Research Office (STIR) and German Science Foundation (DFG). BMGs were selected as suitable material owing to their very high hardness. The material was changed from previously reported reinforcements to 80% $V_f$ W wire (254 μm Ø), 50% $V_f$ W particles (avg. Ø 100 μm) and 50 $V_f$ mixed W/rhenium (avg. Ø 50 μm (W) and 10 μm (Re) particles while matrix remained the same ($Zr_{57}Nb_5Al_{10}Cu_{15.4}Ni_{12.6}$ (Vit106)). Dynamic testing was performed in the form of ballistic tests by firing BMG KEPs onto Al 6061 T651 and 4130 steel targets. Results indicated a new "self-sharpening" behavior due to localized shear band failure. Particulate composite consisting of Zr BMG/50 $V_f$ (W/Re) particles exhibited 8% plastic elongation when tested in quasistatic compression. Strain to failure of 80% $V_f$ W wire/Zr BMG composite with wire aligned with loading axis exceeded 16%. It was more than six times higher than for Monolithic BMG alone. Young's modulus of W wire composite was four times greater than unreinforced matrix material and was in good agreement with RoM prediction. The penetration performance of KEP was 10–20% better than that of W heavy alloy penetrator alone. The study was further continued in 2002 [19] and Vit106 was reinforced with 50 $V_f$ of Mo, Nb and Ta particles. A detailed characterization based on XRD, DSC, EPMA and SEM was performed. The composites were tested in compression and tension. Compressive strain-to-failure increased up to a factor of 12 compared to unreinforced Vit106 BMG. This was due to restriction offered by particles to shear band propagation and generation of further shear band as a consequence. The detail of the study and experimentation could be found here [19]. Based on these studies, further investigations in the field of ex situ BMGMC picked up interest of various research groups in the world. For example, a research based on studying the relationship between flowing velocities and processing parameters such as pressure, $V_f$ of fibers and infiltration length was carried out at UST, Beijing in 2006 [283]. The matrix used was $Zr_{47}Ti_{13}Cu_{11}Ni_{10}Be_{16}Nb_3$ (Vitreloy$^{TM}$1) and reinforcements were W particles. Experimental measurements of contact angle and surface tension of BMG on W was conducted using sessile drop technique at temperature of 1,023 and 1,323 K with different processing time. Wetting angle between melt and W was attributed to belong to reactive wetting as a result of observation of diffusion band at the fringe of metallic drop. This was first good contribution toward studying the chemistry (surface tension and wetting angle) behavior of BMGMCs. This triggered collaborative work between the United States and China on these classes of materials and various studies aimed at studying different processing and structural parameters of BMGs such as effect of hydrostatic extrusion on porous W/Zr BMG melt-infiltrated composites (resulting in much improved fracture strength (1,852 MPa) due to stable interface between MG and W phase and high dislocation density of W phase) [284] and effect of temperature on quasistatic and dynamic testing [285] of W/BMGMCs were conducted. Some modeling studies were also conducted by notable research groups [286] aiming at predicting the selection of composite based on plasticity rather than brittleness. Their details could be found in citations of original work in 2006. During the same time, Professor Schuh at MIT in collaboration with their Singapore–MIT alliance and NUS, Singapore, conducted a detailed dynamic study of different dispersion in situ added to Zr-based BMG alloys [287]. They studied the effect of strain rates, temperature and different levels of volume fraction ($V_f$) of Al (precipitated as Al intermetallics (IMC)) as reinforcements in $Zr_{49}Cu_{(51-x)}Al_x$ ($x$ = 6, 8, 10 or 12 at%) alloys. All experiments exhibited that mechanical properties are dominated by deformation of the amorphous matrix phase. This includes inhomogeneous flow and fracture at low temperatures and homogeneous flow of both Newtonian and non-Newtonian nature at high temperatures. This strengthening effect (in homogeneous deformation range) caused by Al particles increase as $V_f$ increase and was quantitatively explained in both the Newtonian and non-Newtonian regimes. A shear transformation zone (STZ) model developed previously by Prof. Ali S. Argon [288] was also used to describe the behavior and was coupled with model for reinforced particles. This was found to arise from: (i) load transfer from the amorphous matrix to the reinforcements; and (ii) a shift in the glass structure and properties upon precipitation of the reinforcements. An additional mechanism termed as "in situ precipitation" during deformation was held responsible for the increase of strength. This was first contribution of its kind to the field of BMGMCs by MIT, its alliance and collaborators under US Army Research Office. Research is still going on, on this class of material and few notable studies in recent time on this topic have been reported [289–293]. Much attention has been focused on studying their so-called in situ behavior as exemplified by surface tension, surface chemistry, hydrophilic and hydrophobic response and wetting angle studies [291, 292] as well as increase of "plasticity" as predicted by both experimental (static and dynamic tensile and compressive) and simulation studies [31, 222, 225, 293].

### *b.* In situ formation

Research from ex situ reinforced composites shifted to so-called in situ reinforced composite due to increased plasticity observed in this class of materials. The impulse for this was served not only by the mechanism (chemical reaction) but also by the ease with which these materials are fabricated coupled with observation of enhanced properties (plasticity/ductility) by single step process (i.e., NG of dendrites directly from melt during solidification) rather than external mixing (mechanical or physical (infiltration)) of reinforcements which not only is difficult but is energy extensive process adding an additional cost to process. The interest was also triggered by the fact the established mechanisms and theories of solidification (NG) could be applied to increase the properties just like conventional metallurgical alloys and that this phenomenon was observed in other very important class of materials, that is, Ti-based BMGMC with much superior casting and mechanical properties. Although research triggered by synchrotron light later on revealed more complex mechanisms of formation (liquid state phase separation prior NG), still, this method of production of BMGMC dominate, by far the amount of research activity focused on these class of materials. Investigations such as by high-energy radiation and simulations [294, 295] just helped to further polish the mechanisms, processes and/or improve properties of composites.

Credit again goes to Caltech and Ms. Haein Choi-Yim for introducing this concept first time in 1997 during her PhD studies [18, 296], but the real effort came from more permanent members of group Hays in 2000 [12] when they first reported the use of fine dispersed ductile phase crystalline (bcc Ti-Zr-Nb β phase) dendrites in Ti-based BMG matrix (Figure 8 (a)) to increase plasticity and study and control their microstructural properties. This was indeed major discovery as 3D network of equi-axed dendrites produced as a result of controlled solidification and micro alloying effectively not only served as barrier to shear band propagation but also become the reason for their multiplication. This dramatically increased the plasticity, impact resistance and toughness of BMGMC.

They improved on their discovery in 2001 [297] and it was shown that shear band spacing is coherent with periodicity of dendrites. Mechanical tests were conducted under unconstrained conditions. The generation of shear bands, their propagation and interaction with dendrites is attributed to "free volume" concept rather than dislocation theories which are major mechanism of failure in crystalline alloys. At the same time, another group at Caltech studied the effect of shock wave on Zr-based BMG and their newly formed composites [298]. They subjected conventional

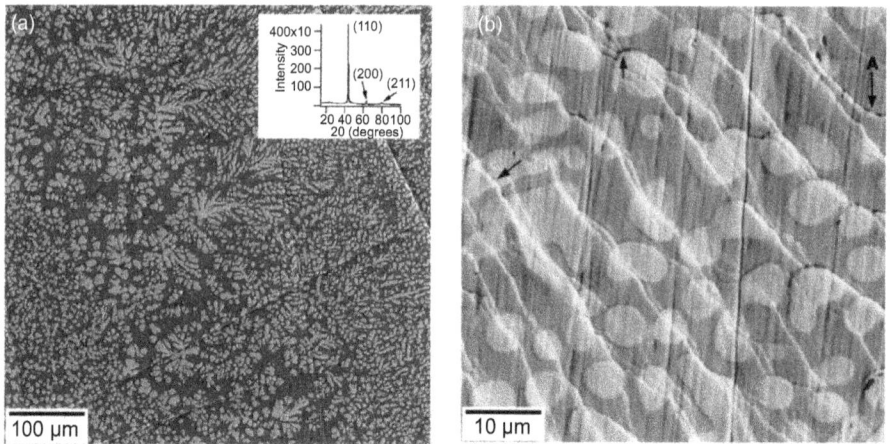

**Figure 8:** (a) SEM backscattered electron image of in situ composite microstructure (×200) and (b) shear band pattern array from failed surface showing their crossing dendrites [12].

Vitreloy$^{TM}$1 and new β-Vitreloy$^{TM}$1 to planar impact loading. They observed a unique low-amplitude elastic precursor and bulk wave prior to rate-dependent large deformation shock wave. Spalling was also observed in both Vit1 and β-Vit1 and was attributed to shear localization and debonding of β phase boundary from matrix at 2.35 and 2.11 GPa (strain rate $2 \times 10^6$ s$^{-1}$), respectively. Since then, the interest in development and improvement of these composites has increased many folds in various groups around the globe. As said earlier, special focus is given to improvement of metallurgical properties related to thermodynamics and foundry engineering. A notable contribution in this regard is made by a group led by Prof. Jürgen Eckert at IFW Dresden. He and his colleagues have studied various properties of in situ BMG composites from various perspectives related to processing, casting [299–301], corrosion and mechanical. For example, in a study in 2004 [299] one of their graduate students (Mitarbeiter), Jayanta Das (now at IIT Kharagpur) studied the effect of casting conditions on mechanical properties of Zr-Cu-Al-Ni-Nb alloys with and without the addition of Nb. It was reported previously that the addition of Nb strongly affects the casting properties (fluidity) of this otherwise sluggish and viscous alloy. The research showed that addition of Nb caused a stabilized bimodal size distribution which consisted of bcc β-Zr nanocrystalline dendrites dispersed in amorphous matrix. It was observed that a strong relation exists between the dendrite and matrix crystallite size and morphology which is a function of casting conditions. The cooling rates ($\varepsilon \sim 2.6 \times 10^3$ to $4.0 \times 10^1$ K/s) were estimated from secondary dendrite arm spacing.

Effect of processing parameters on the mechanical properties is less significant for the lower percentages of Zr and Nb which was increasingly significant for $Zr_{73.5}Nb_9Cu_7Ni_1Al_{9.5}$. Similarly, optimization of properties (reaching maximum of

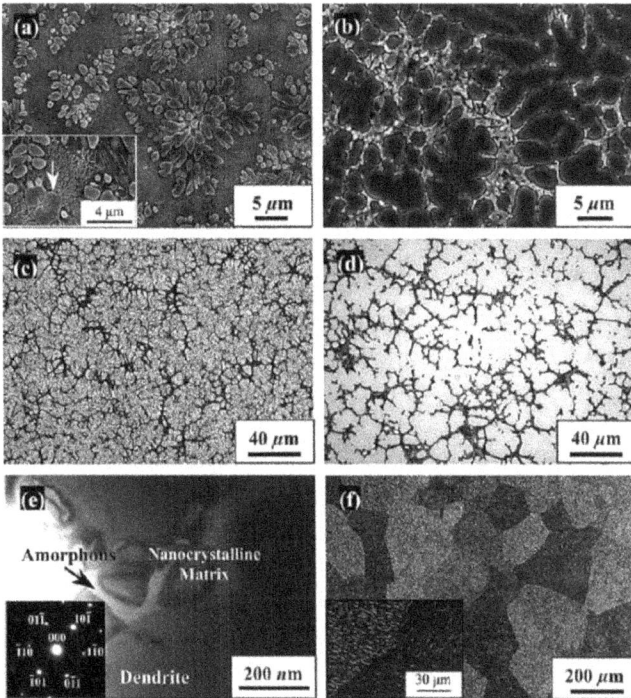

**Figure 9:** Typical microstructures of Zr-based in situ dendrite/glass matrix reinforced composites. (a) Injection cast rod (Ø 5 mm) (inset: magnified view of IMC), (b) suction cast rod (Ø 5 mm), (c) centrifugally cast rod (Ø 5 mm), (d) cold crucible cast rod (Ø 10 mm), (e) TEM bright field image of centrifugally cast rod (Ø 5 mm) showing dendrites (inset: SEAD of dendrite) and (f) SEM image of cold crucible cast rod showing various grains (inset: magnified view of precipitates within grains).

1,754 MPa fracture stress and 17.5% strain to failure) was possible only toward a rod of Ø 10 mm. Another very important fundamental discovery was the observation of dislocations in these alloys. Although it was observed in dendrites (while shear banding was dominant in bulk glassy matrix), it still was major breakthrough toward increasing plasticity of these BMGMC. Figure 9 shows optical and electron micrographs of Zr-based in situ dendrite/glass matrix composites cast in the form of rods of Ø 5 mm and Ø 10 mm (cold crucible only) by various manufacturing routes. Another important development which took place in their group was the introduction of transformation-induced plasticity (TRIP) BMGMCs. They successfully produced BMGMC in Martensitic alloys [302] and also caused transformation of B2 ductile phase (observable in Zr-Cu-Co systems) to B19´ martensitic phase (with or without twins) [21, 303–305] in this class of material. They are also pioneer of (not quite matured) production of BMGMC by selective laser melting (SLM) (a form of AM) [59, 306, 307] (described in detail in Section 3.3.1.2). Since then various studies have been reported by their group not only in Zr but Al- [308], Fe- and Ti-based

[274, 277, 309, 310] BMGMCs in numerous reports [23, 35, 50, 246, 256, 295, 302, 303, 306, 309, 311–317] showing improvements in metallurgical, chemical and mechanical properties of these composites. They are heavily supported by clusters such DFG, DAAD, EU Marie Curie network, FP7 and Humboldt Foundation and have strong collaborative programs with Caltech, JHU, UTK, UST, Tohoku and IITs.

Another group whose contribution toward development and processing of this class of materials is worth mentioning is led by Prof. Todd C. Hufnagel at Johns Hopkins University. They also contributed by fundamental physics [318], microstructural [32, 319] and mechanical property [203, 320, 321] improvement. Their contributions led to development of increased plasticity and improved fracture toughness by conventional manufacturing processes. One of the notable discoveries made by their group details the deformation of BMG by slip avalanches [318]. In 2014, Professor Hufnagel along with his collaborators for the first time reported that slowly compressed BMG deform via a slip avalanche of elastically coupled STZ which are a collection of 10–100 atoms. It was previously believed that deformation in BMG is largely due to rapid movement of shear bands *(which is intermittent instantaneous transience (rearrangement) of atoms)*. However, the detailed mechanism about their behavior at low and high temperatures and slow and high deformation rates was not known [3]. In their study, they reported that slip avalanches are observed at very slow compressive deformation rates only in accordance with serrated flow behavior. A schematic of the setup used and patterns observed are shown in Figure 10. At very high temperatures, there is no problem in explaining the behavior of BMGs which is primarily homogeneous. However, this does not remain same at low temperatures and changes to the inhomogeneous type which is characterized by onset of intermittent slips on top of narrow shear bands. If this inhomogeneous deformation is coupled with low strain rates, it gives rise to sudden stress drops called serrated flow which was found to be consistent with earlier reports [314, 322, 323] (which is new discovery). This discovery led to better understanding of behavior of BMGs and consequently provided basis for the manufacture of strong and tough materials.

Elaborative details of their work could be found in reports published from their group [3, 203, 294, 318, 319, 321, 324–327]. There are further more notable contributions from the United States (e.g., Prof. Peter K. Liaw at UTK (primarily about fatigue, shearing and nanoindentation) [41, 239, 328–341], Prof. Jan Schroers at Yale), Taiwan [10, 342] and Korea [135, 257, 343–345] which provided footstone for the further development and property enhancement of this class of materials.

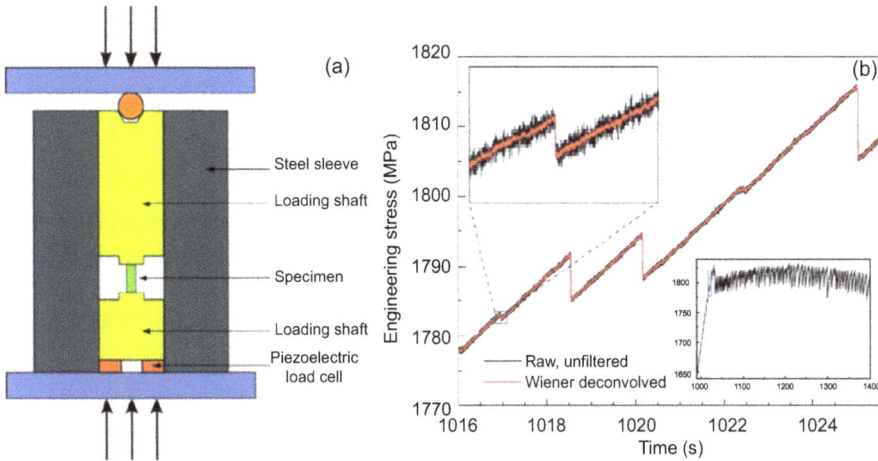

**Figure 10:** (Left) Schematic diagram of experimental setup. (Right) lower right inset: stress versus time, main figure: magnification of data in small window in lower right inset. Upper left inset: magnification of stress drop (serration) during slip avalanche (black lines: raw unfiltered stress versus time; red lines: time series after Wiener deconvolution) [318].

### 2.8.2 Solid–solid (S–S) transformation (devitrification)

A remarkable property exhibited by BMGMC is that after being casted in vitrified shape, they tend to relax when subjected to the effect of heat. This is very much similar to phenomena observed in monolithic BMG which accounts for their ductility. This gives rise to another mechanism by which they could be formed with enhanced strength and toughness known as devitrification or solid state phase transformation. This is primarily attributed to either of two mechanisms which may be responsible for the onset of crystallization upon heating. These are known as "solid state phase separation" [41] and "quenched in nuclei" [37, 44]. Solid state phase separation also known as "spinodal decomposition" in conventional alloys [346] is separation of phases which occurs before the growth of second phase in BMGMCs. This is not NG phenomena but solid state detachment based on miscibility of two phases with each other. Although observed in selected compositions only [257, 260, 264, 265, 347, 348], this accounts for a unique mechanism for the formation of ductile dendrite reinforced BMGMCs [262, 349]. Second mechanism which is known as "quenched in nuclei" are intermittent metastable nuclei of ductile phase dendrites which have just started to grow but their growth is suppressed by rapid cooling to which they are subjected quickly thereafter. This retards any further growth. However, as they are metastable and still in their high energy state with the drive to transform to fully grown stable phases, they fulfil their desire when subjected to heat and serve as sites for the growth of full-scale second-phase crystalline dendrite. This is very important mechanism of fabricating these kinds of composites. These are rather well-established mechanisms and account well for the phenomena observed.

## 2.9 Production method – process perspective

From a process point of view, production methods could be broadly classified into five main methods used for the fabrication of BMGMCs till date: melt spun, twin-roll casting (TRC), Cu mold suction casting, semisolid processing and AM. This classification by any means is not exhaustive but narrates only those methods which are easy, cheap, reproducible, versatile and have been used repeatedly over years for the manufacture of ductile glassy composites. Their brief details are as follows:

### 2.9.1 Twin-roll casting

This, by far, is the first method ever used for the manufacture of glassy alloys. This is a variant of melt spun method which is responsible for the first observations of glassy structure in metals by Paul Duwez and coworkers [1] as they casted thin ribbons of Au-Si alloy which were first to retain glassy structure at room temperature. The process in essence consists of two high-speed opposite direction rotating self-driven water-cooled Cu rolls, which are finely ground and polished to avoid any contamination and variation in thermal gradient across surface caused by impurities. Furthermore, the wheels are constantly replenished with lubricating medium (usually organic surfactants) (if necessary) to ensure their smooth functioning and near friction free rotation. This and other processes have been excellently documented by Professor Inoue in a recent article [350].

TRC underwent a considerable hierarchical development from an earlier variant of wire caster to rotating disk caster finally in its TRC form (Figure 11).

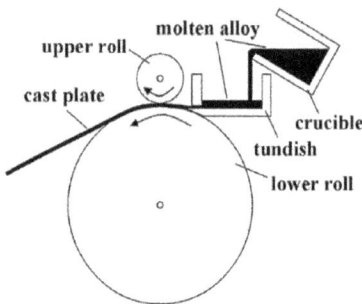

**Figure 11:** Schematic twin-roll caster (TRC) setup [350].

Modeling and simulation of the process and the solidification microstructure evolved in it has also been carried out [351] to further optimize the processing conditions and set process parameters. The process is excellent for making thin continuous lengths of ribbons of almost all compositions. Since process is not affected by any change in alloy composition. Chilling effect is caused by the very nature of water cooled copper rolls, their surface area and rotation. Water temperature is usually kept at 8 °C and speed of roll is variable (could be adjusted for maximum efficiency). Copper is selected as a material for rolls as it has excellent thermal conductivity; thus, it can easily conduct heat away from it. Alloy is poured from ladle or crucible to a tundish which transfer the melt to rotating copper rolls. It is by far the first and still the only mass production method for making BMG and their composites.

### 2.9.2 Cu mold (suction) casting

As the conductivity of copper is maximum and it is very good material for heat transfer, it paved the way toward an idea of using it as mold to make ingots rather than ribbons. The chilling effect is caused by water cooling coupled with suction caused by difference of pressure at the bottom of mold (creation and destruction of vacuum) to that at top. This results in bulk shapes of BMGs. This was, and still is, the most frequently used method for the production of 3D BGA. The process essentially is the melting of BMG in a small arc melting button furnace. This is followed up by the introduction of melt over a nozzle-type opening which is controlled by a suction duct/chute attached to a vacuum pump at the bottom. The pipe connected to vacuum pump is attached to inert gas (Ar) cylinder at the same time with the help of three-way rotating valve. Once the alloy is in its molten state over suction chute (nozzle), three-way valve is quickly shifted toward vacuum pump which sucks all the argon present in chamber through that chute thus takes away the alloy with it as well. The alloy along its way toward vacuum pump encounters water cooled chilled Cu mold (plate or cylinder) thus quickly gets solidified just like any conventional casting process. Once casting is completed, the valve is turned toward air and set up is allowed to cool a little before opening (mold break out/fettling). This is very efficient, quick method for the production of BMG and their composites and again nearly all types of compositions could be casted as process is composition independent. There have been different approaches the way different groups have been doing the castings using this technique [352–357] but the fundamental principle remains the same. A recent approach adopted at Huazhong UST, Wuhan [358], is described in Figure 12.

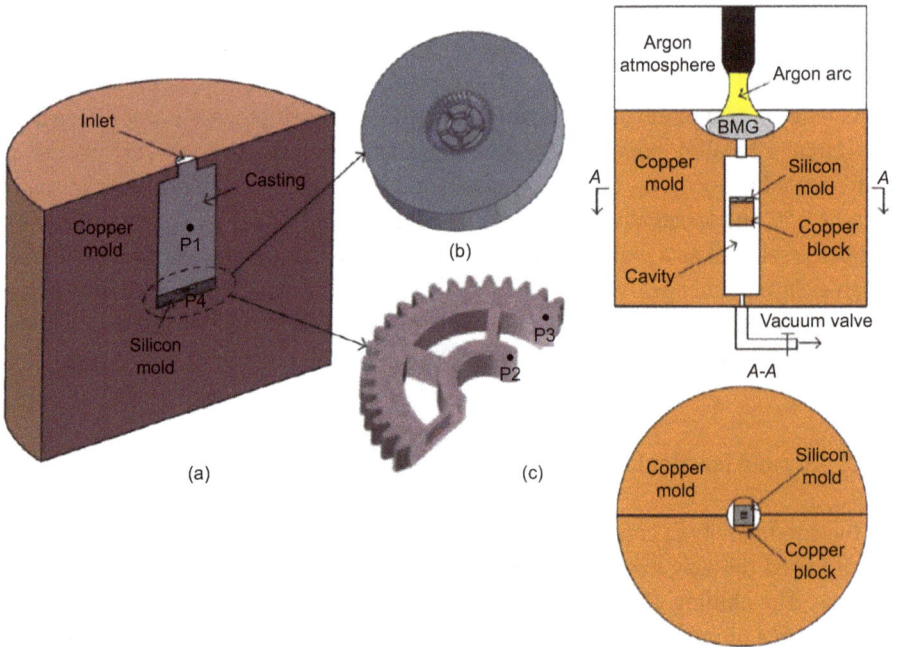

**Figure 12:** (Left) (a) 3D view of casting assembly (b) and (c) expanded view of silicon mold for casting of gear (right) (upper) 2D side and (bottom) top view of micromold for suction casting [358].

### 2.9.3 Semisolid processing

Semisolid processing, most popularly tried in its variants forms such as "semisolid forging," is another innovative technique for processing BMGMCs. Essentially, in this process, alloy is melted by taking the composition to its melting temperature and then upon cooling first it is held in a region in between liquidus and solidus lines in which NG of second-phase particles (dendrites) is fully completed. Following this, the alloy is rapidly quenched to room temperature. This "hold" in between liquidus and solidus ensures the formation and homogeneous distribution of ductile phase equiaxed dendrites which are necessary for the development of ductility and toughness. The process is unique in a sense that alloy stays in semisolid state when it is going through primary liquid-to-solid transformation (solidification). The hardness of material in this region is low and it is possible to impart external forces on it causing its deformation. This is called "semisolid" processing. From practical view point, it is the only method of fabrication of semi complex shapes from BMGMCs. Researchers at Caltech in parallel with UC San Diego in 2009 [359, 360] used this technique to devise a novel method employing "double boat" and fabricated hemispherical shapes out of different compositions. This was by the application of force using a stub (named as forging). The process is described in detail in Figure 13(a–d).

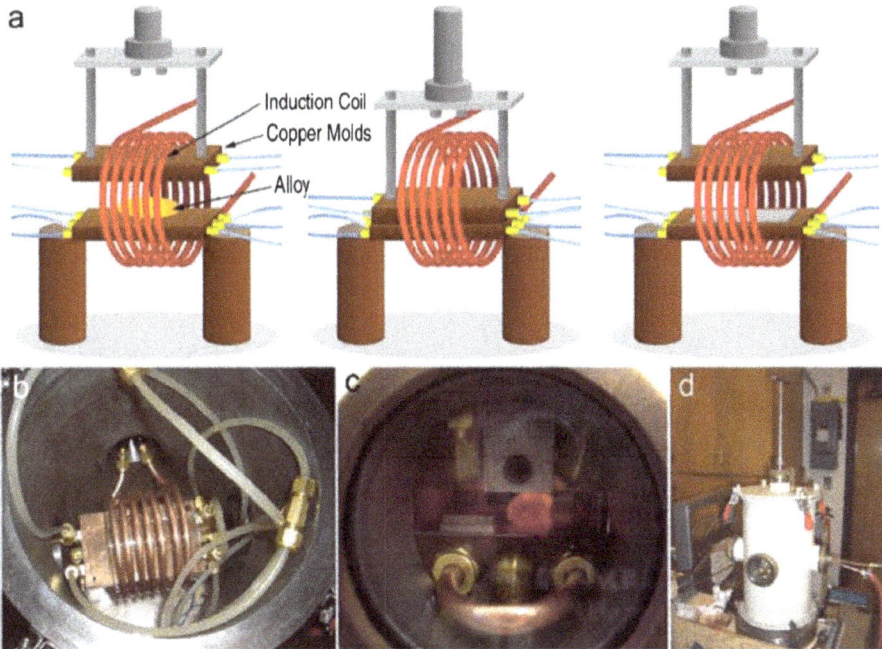

**Figure 13:** (a) Schematic of double boat method, (b) photographs of mold together inside the casting chamber, (c) casting chamber closed and (d) external full view of setup showing inlets and cooling water [359].

The same technique and its extended version was used for microgravity experiments on board NASA Space Shuttle Program during the 1990s in three notable missions: STS–65, STS–83 and STS–94 [361]. There were other efforts following this as well which were based on same principle, but the technique was extended to make complex shape by so-called super plastic forming [362] or "TPF (blowing [60, 216])" [10, 215, 217, 362–364].

## 2.9.4 High-pressure die casting

High-pressure die casting (HPDC) of MG is recently reattempted. Fe- [365] and Zr-based [366, 367] MG have been formed by HPDC. First time, it was attempted in 1992–93 [368] when $Mg_{80}Cu_{10}Y_{10}$ [369] and $La_{55}Al_{25}Cu_{10}Ni_{10}$ [370] are casted in small cylinder and sheet form with diameters or thickness ranging from 1 to 7 mm. Recently, in Fe-based alloys, a HPDC tool (key shaped) is designed for $Fe_{74}Mo_4P_{10}C_{7.5}B_{2.5}Si_2$ [365]. This composition was chosen because of its relatively low melting point, good GFA and good soft magnetic properties. Another advantage is its low affinity toward oxygen. Key-shaped specimens of maximum and minimum widths of 25.4 and 5 mm are produced, respectively. The influence of die material, alloy temperature and flow rate on microstructure, thermal stability and soft ferromagnetic properties have been studied. The results suggest that a steel die in which the molten metal flows at low rate and high temperature can be used to produce completely glassy samples. Detailed analysis involving study of heat transfer coefficients and flow pattern indicates avoidance of skin effect which is augmented by proper tool geometry. Magnetic measurements also indicate that amorphous structure of the material is maintained throughout the key-shaped samples. This study reveals that HPDC is an excellent method for casting of Fe-based BGA even with complex geometries.

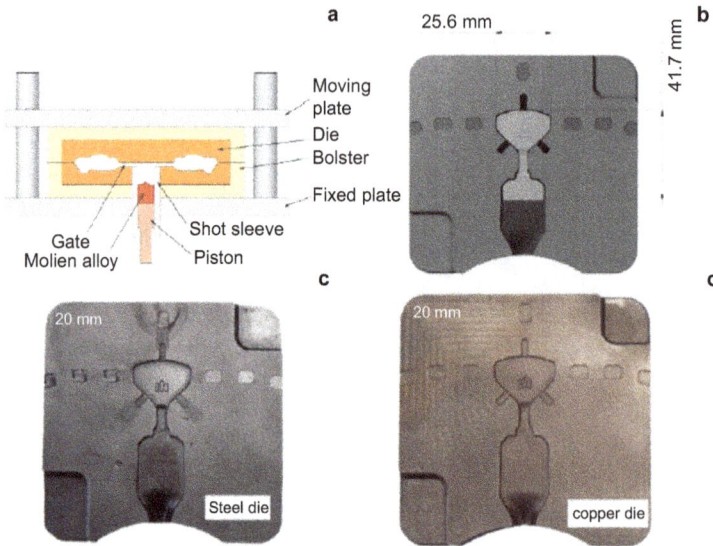

**Figure 14:** (a) Schematic illustration of the high-pressure die casting setup, (b) 3D model of the die and (c) completed dies made from heat-resistant steels and a copper alloy [365].

Similarly, forming ability of HPDC is determined for Zr-based MG ($Zr_{55}Cu_{30}Ni_5Al_{10}$) [367]. Alloy is formed by addition of trace of rare earth metal yttrium. Study determines the critical size of forming BMG, for design and fabrication of BMG components by HPDC. It indicates increasing applied pressure during casting suppresses the nucleation of crystal nuclei but does not affect their growth at a deep supercooled temperature. This decreases critical cooling rate and increase critical size. This critical size is determined to be about 4–7 mm under different HPDC parameters. In another study, HPDC is used to form near net complex shapes [366]. This study utilizes entire process vacuum HPDC to fill die cavity in milliseconds and create solidification under high pressure. This compensates for lack of time of the window for shaping under required high cooling rate. This technique allows glassy structure to be produced for most Zr-based BMGs with a size of 3–10 mm and high strength.

### 2.9.5 Continuous casting

Continuous casting is a process in which continuous lengths of material can be casted by pouring liquid metal at one end and allowing the solidifying metal to exit from other by passing from a mold [371]. The geometry of mold dictates the shape of casting and can be controlled by various system and process variables. Resulting product may be obtained in the form of plate, slab, rod or bar. These may be cut to specific dimensions on exit side by using oxyacetylene flame but the process of production of shape never stops [107]. This process has also been applied to BMG [372] and continuous lengths are produced taking advantage of process such excellent dimensional control, material saving, continuous nature and shape control. Both experimental and simulation studies of the process revealed that maximum cooling rates achievable with the setup are 15–17 K/s which is well above the critical cooling rates for glass formation of $Pd_{43}Ni_{10}Cu_{27}P_{20}$ but below $Pt_{57.3}Cu_{14.6}Ni_{5.3}P_{22.8}$ and indicates crystallization in later case while amorphization in the former. In another study vertical-type TRC is assessed as process for continuous casting [373]. Temperature and flow phenomena of melt in molten pool were calculated. It is determined that cooling rates play a very important role. Material derivative method based on continuum theory is adopted in cooing rate calculations. TTT and CCT diagrams are calculated.

It is found that vertical TRC has the good ability to produce continuously cast glass ribbons.

### 2.9.6 Squeeze casting

Squeeze casting is another new yet relatively less used process for production of BMG [374]. In this, liquid metal is pressure fed into mold cavity placed in between plates of hydraulic press and allowed to solidify there. Mold cavity serves wall of die which is usually water cooled. A combination of water cooling and pressure produces a rapid heat transfer condition which results in pore-free structure. It is applied to $Mg_{65}Cu_{15}Y_{10}Ag_{10}$ MG. Alloy is melted in vacuum induction melting furnace from where it is bottom filled into a water-cooled Cu mold (ø 10 mm, L 75 mm. After filling pressure is raised to and maintained at 100 MPa for 120 s (2 min)). The resulting product is mostly defect free and has uniform grain structure. Two phenomena dominate during the process: (a) increase of melting point of alloy [375] as per the Clausius–Clapeyron equation:

$$\frac{\Delta T}{\Delta P} = \frac{T_f(V_1 - V_s)}{\Delta H_f}$$

where $T_f$ is equilibrium freezing temperature, $V_1$ and $V_f$ are specific volume of liquid and solid respectively and $\Delta H_f$ is latent heat of fusion (constant) and (b) rapid rate

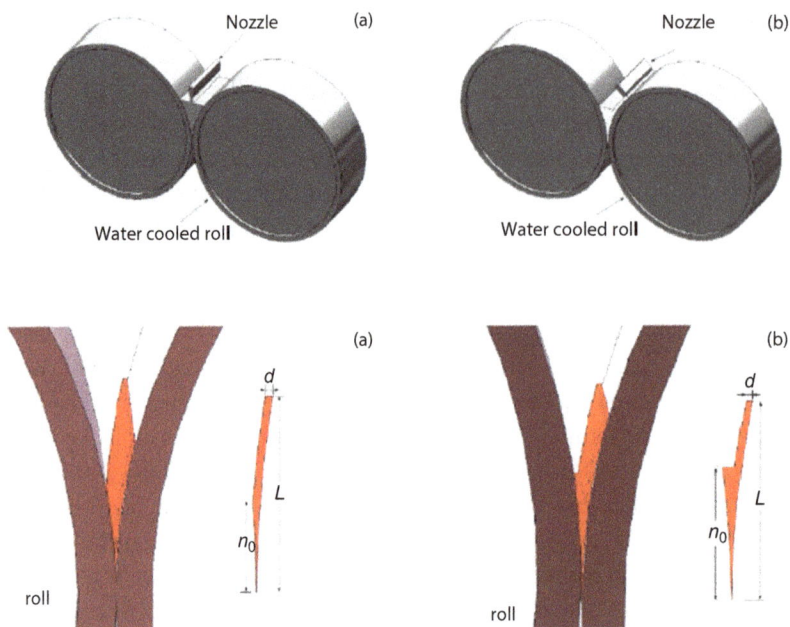

**Figure 15:** (Top) Schematic of twin-roll casting process, (a) center pouring mode, (b) One side pouring mode, (bottom) schematic of molten pool of one side pouring mode (a) $h_a$ depends on roll gap and $d$ and (b) $h_o > h_a$ and can change independently.

of heat transfer from walls of die as hot metal encounters cold walls. This results in large undercooling which promotes glassy structure formation.

### 2.9.7 Injection molding

Owing to excellent TPF properties of certain BMG compositions, they can be easily fabricated into intricate parts using injection molding technique. For example, $Zr_{35}Ti_{30}Be_{27.5}Cu_{7.5}$ is one such alloy. Its crystallization glass transition temperature is 165 °C [376]. Its large supercooled liquid region (SLR) provides the longest processing times and lowest processing viscosities of any MG known, making it suitable to be processed by injection molding. Some other efforts were also made making use of superior TPF properties of certain composition of BMG and satisfactory results are obtained [377]. This made themselves an application in micro optics with high volume fabrication [378] and nanoinjection molding [379].

## 2.9.8 Centrifugal casting

Centrifugal casting of BMG has also been performed recently [380]. Metallic glass samples in the form of rings have been produced with an outer diameter of 25 mm and controlled thickness (>1 mm) by changing the weight of molten alloy. This is a very useful method for production of rings, tubes and cylinders of varying thickness and lengths. The process has inherent advantages of being versatile, flexible, having sharp and continuous temperature gradient which allows rapid solidification necessary for glass formation and easy to control. In present study, molten metal produced by vacuum induction melting in protected argon atmosphere is injected into a cylindrical copper mold rotating at a speed of 2,500, 3,000 and 3,500 rpm. Copper mold has the following dimensions:

$$\emptyset_o = 150 \text{ mm, } \emptyset_i = 25 \text{ mm and thickness} = 50 \text{ mm}$$

$Zr_{55}Cu_{30}Al_{10}Ni_5$ is cast in the form rings of $\emptyset = 25$ mm and th $= 1$ mm. The machine employed has the following electrical ratings: voltage $= 380$ V, $f = 50$ Hz, engine power $= 2.2$ kW. It is a vertically aligned machine while the product is produced in a horizontal form as casting chamber rotates about its vertical axis.

**Figure 16:** $Zr_{55}Cu_{30}Al_{10}Ni_5$ rings made by centrifugal casting method.

## 2.9.9 Friction stir welding

Friction stir welding of BMG is a relatively new technique and few studies have been reported encompassing description of joining of BMGs by friction stir welding [381–384]. Using this technique, not only BMGs are welded together [383, 385, 386] but they are also welded to other crystalline metals and materials such as pure Al [387, 388], pure Cu [389] and Al alloys [390]. For example, in a study [385], $Zr_{55}Cu_{30}Al_{10}Ni_5$ plate was successfully friction stir welded below its crystallization temperature. A wider tool (shoulder ø = 25 mm, probe ø = 5 mm and probe length = 1–2 mm) and low angle of recessed shoulder (3°) was used to avoid flash formation and minimize heat concentration at the shoulder edge. DSC, XRD, TEM and micro-hardness analysis confirmed that the amorphous structure and original mechanical properties were maintained in the whole joints. Similarly, in another study, $Zr_{58.5}Nb_{2.8}Cu_{15.6}Ni_{12.8}Al_{10.3}$ (Vitreloy 106a) is welded with fixed polycrystalline cubic boron nitride pin tool below its crystallization temperature. It was observed that the SRO was reduced for the domain size while it was increased for the weld nugget. In another study [387], 2 mm thick $Zr_{46}Cu_{46}Al_{8}$ (BMG) plate and 4.5 mm thick pure Al plate were successfully joined by FSW, and the defect-free BMG/Al joint is produced. Some physical stirring of BMG particles into Al is observed with a reaction product (Al-rich phase) but no crystallization is observed at the interface. In the case of Al alloys [390], 1.7 mm thick plate of BMG was welded to the Al plate. No reaction product or crystallization is observed, and good joint is obtained. Similar behavior is observed for weld joint of BMG to pure Cu in a plate size of 2 mm. No reaction product, layer at the interface or at the particles or crystallization is observed. UTS of joint was about 253 MPa as compared to 432 MPa in the case of BMG with Al-Zn-Mg-Cu and 190 MPa in the case of BMG with pure Al which is quite characteristic.

### 2.9.10 Joining

Joining of BMG may be divided into soldering, brazing and welding (including explosive welding [391]) depending on operating temperature, working temperature and range. Significant advancements have been made in these processes for BMGs [392].

### 2.9.10.1 Thick wall cylindrical explosion

This technique has been extensively applied for joining Zr-based BMG to Cu [393] and Cu alloys (brass). In case of BMG (Vitreloy 1) joining with brass [394] a very good bond is obtained owing to significant atomic diffusion across the welding interface and shock wave propagation in the weldment. Investigation of explosive welding of $Zr_{53}Cu_{35}Al_{12}$ BMG with crystalline Cu [393] revealed that both materials are tightly joined to each other without visible defects and a thin diffusion layer is formed at the interface. Mixed amorphous and partially crystallized structure is observed in diffusion layer but BMG in close proximity to interface still retains its amorphous state. A variant of explosive welding, known as underwater explosive welding [395], has also been applied to BMG and their joining with metals successfully.

### 2.9.10.2 Soldering

Soldering is a process whereby two different materials are joined together by the help of low melting filler metal (such as Sn or Zn). In BMG, it presents a significant challenge to prevent them from crystallization. A novel solution to it has been found by use of ultrasonic waves forming ultrasonic assisted soldering method [396, 397]. In one example, Fe-based BMG is joined using this method [396] while Zr-based BMG is tested in other case [397]. No obvious defects were observed, and the joining ratio was more than 90% in the former case. At the interface of solder/Al, scallop α-Al solid solutions were formed and large 50 μm fretting (pits) were found on Al plates. At solder/amorphous interface, $FeZn_{13}$ intermetallic was formed. Adding nickel to Sn–Zn solder can improve the hardness of the joints by formation of $NiAl_3$ intermetallic particles and depressing Al corrosion. Same phenomena when applied on Zr base BMG is best explained by sonocapillary effect. This is collapse of cavitation bubbles on BMG surface causing extreme erosion. This increases the adhesion properties of solders.

### 2.9.10.3 Brazing

The term *brazing* is used to describe three phenomena in BMG, that is, brazing method or process, brazing filler [398–401] and brazing with a certain technique such as electron beam melting [402] and vacuum [400]. In first case [399, 403, 404], $Cu_{54}Ni_6Zr_{22}Ti_{18}$ BMG was brazed to SS400 carbon steel using (pure Zn and Zn–Ag–Al) filler metal at 713 K under a compressive stress of 15 MPa for 3 min. These filler metals suppress the crystallization during a reheating thermal cycle, forming a sound bonding interface without notable chemical diffusion particularly with the Zn–Ag–Al filler resulting in shear bonding strength of 45 MPa. Owing to low bonding temperature and short bonding time, the present method marked no change in the phase composition of the BMG. Slight variations in glass transition temperature and crystallization temperature are due to the thermal annealing effect. Similar studies are carried out on another alloy TiCuZrPd on Ti-6Al-4 V substrate [405]. BMGs are extensively used as filler metals. For example, $Ti_{33}Zr_{17}Cu_{50}$ is used as filler in vacuum at 1,123–1,273 K to join $ZrO_2$ and Ti6Al4V. It is demonstrated that microstructure and mechanical properties are significantly affected by brazing temperature, heat time and cooling rate. These are 1,173 K, 10 min and 5 K/min in this case. A brazing seam consists of $ZrO_2$/$Cu_2Ti_4O$, $(Ti,Zr)_2Cu$/TiO, $Ti_2O$/$CuTi_2$, $(Ti,Zr)_2Cu$/$CuTi_2$/Ti–6Al–4 V.

### 2.9.10.4 Welding

Various types of welding (laser [76–78, 80, 81, 406–415], electron beam [392, 416], ultrasonic [417–421], friction [422–424], friction stir [381–386, 389, 390, 419, 425–428], explosive [391–393, 395, 429, 430], resistance spot [431–433] and GTAW [105, 106, 434]) are used for joining BMG to each other and with other metals and materials. Notable among these are gas tungsten arc welding and resistance spot welding as others have already been explained in previous sections. Both these processes are essentially conventional processes which are modified to account for BMGs by change of certain variables such as type of gas, its flow rate, source, machinery (holder, nozzles and W electrodes), feed rate and its control in case of former and spot temperature, spot size, sheet thickness which may be weldable in case of later. Their careful control results in refined microstructure (fine equiaxed grains, size and distribution of particles, low or no porosity, no shrinkage, no hot spots, no overlaps, refined heat affected zone, and 100% fused fusion or melt zone) and improved mechanical properties such as YS, UTS, percent elongation, hardness, impact, fatigue and creep properties. Their individual detail is omitted in the interest of time. Interested reader is referred to the cited literature.

### 2.9.11 Foams

MG foams are known as early as 2003 when $Pd_{43}Ni_{10}Cu_{27}P_{20}$ is researched at Caltech [435]. This is mixed with $B_2O_3$ which releases gas at elevated temperature and/or low pressure. This alloy is sluggish even at liquidus temperature which helps in foaming. By raising the temperature to liquidus and decreasing the pressure to $10^{-2}$ mbar, foam densities as low as $1.4 \times 10^3$ kg/m$^3$ were obtained. Bubble diameter ranges from $2 \times 10^{-4}$ to $1 \times 10^{-3}$ m while its volume fraction is at 84%. Other studies suggested similar trends and results [436–439] are described below.

### 2.9.11.1 Syntactic

Syntactic processing involves melt infiltration of allot (in present case $Zr_{57}Nb_5Cu_{15.4}$-$Ni_{12.6}Al_{10}$) into a bed of hollow carbon microsphere followed by quenching [436]. Resulting foam consists of glassy metallic matrix containing ~60 vol% of homogeneously distributed carbon microspheres $\phi = 25$–$50$ μm.

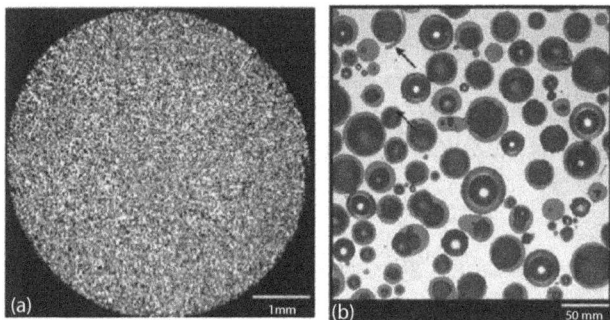

**Figure 17:** Optical micrographs showing structure of syntactic Vit106 foam. (a) Low magnification image and (b) high magnification image.

In another study [440], foam is fabricated from BMG and alumina cenospheres. This new foam possesses greatly enhanced energy absorbing capacity of 113.6 Mj/m$^3$ due to combination of high strength, stability and ductility. It is observed that high strength results from alumina cenospheres. Both the collapse of struts and multiple shear bands in MG matrix accommodate high deformation.

### 2.9.11.2 Ductile

In this type of alloys, a technique known as salt replication technique is used. An open cell fully amorphous metal foam is fabricated from commercial glass forming alloy Vit106 by casting it into a sintered $BaF_2$ (chosen due to its high stability and melting point) using low-pressure melt infiltration followed by rapid quenching. This rapid quenching and leaching is performed in a bath of 2 M $HNO_3$. The resulting alloy is ductile in compression and achieves and engineering strain of 50% without failure.

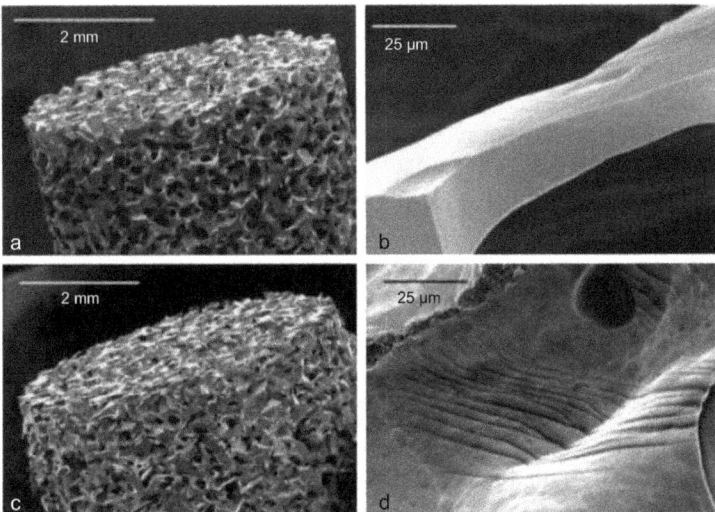

**Figure 18:** SEM images of amorphous Vit106 foam with 22% relative density. (a) Microstructure prior compression, (b) individual strut within undeformed foam, (c) microstructure after 50% compression and (d) deformed strut in compressed foam.

A similar study also suggest analogous results in which powder extrusion is used as method of fabrication [441].

### 2.9.12 Thin films

Thin-film MG are a class of materials which could form thin layer on a substrate. This may serve various purposes such as passivation, protection and inhibition. Their details are explained in many recent excellent studies on the subject [442–446]. They are primarily used for improvement in surface properties of alloys serving as substrates such as corrosion resistance [445, 447–454], fatigue resistance [451, 455–458], electrical properties [449, 459], biocompatibility [445], indentation [460] and scratch resistance, thermodynamic (including thermal resistance) properties. They may be manufactured by vapor quenching [461], bulk quenching [462], magnetron sputtering [463, 464], ion beam sputtering [465], laser processing (deposition) [448] and superplastic forming [466]. Their detailed in-depth analysis and comparison may be obtained elsewhere. Interested reader is referred to the cited literature.

### 2.9.13 Additive manufacturing

This is the most recent and advanced process of making BMGMCs (also the subject of present study). Various efforts have been made to produce Al- [63, 306, 467–469], Fe- [62, 307, 470], Ti- [471–474] and Zr-based [69, 70, 475–480] BMGMCs by different variants (selective laser sintering (SLS), SLM [59]/laser-engineered net shaping (LENS)® [481], powder feed deposition (PFM)/direct laser deposition (DLD) [482, 483], wire feed method [484], electron beam method [67]) of AM [71, 83, 476]. Process essentially consists of supplying material (BMGMC) in the form of powder coaxially or at an angle to vertical axis (PFM/DLD), or wire (WFM) just like welding (MIG/GMAW) or making shape out of powder bed by selectively sintering (SLS)/melting (SLM) parts by the impingement of laser on the bed. The motion of laser is controlled by the robotic arm which is controlled by a microprocessor and this motion is a function of coordinates given to controller (computer) in the form of CAD model. There are variants in which CAD model could be generated as a part of whole set up, is sliced by a slicing program (which divides the geometry of part into layers (layer-by-layer (LBL) fashion)) and is given to final laser controlling robot which dictates its motion as LBL pattern forms (Figure 19). **Advantages** of the process are: (a) it is one-step direct fabrication of composite parts in near-net shape, (b) there is no need for any posttreatment (finishing, deburring and heat treatment) facilitating the attainment of properties directly without incurring additional costs, (c) surface finish of parts produced is excellent, (d) excellent dimensional accuracy could be achieved and finally (f) excellent microstructural control could be exercised in one step. **Limitations** are: (a) it is still a new process and optimization of properties as a function of laser parameters is a challenge, (b) reproducibility is difficult from laboratory to laboratory, (c) almost nothing is known about metallurgy of melt pool due to its incipient and transient nature thus mechanisms of NG coupled with heat transfer are not known (aim of present study), (d) powders for AM require special production procedures and are expensive and (e) process is largely dedicated to repair (including cladding) rather than full-scale part production due to aforementioned reasons.

Notable discoveries have been made by groups at Ohio State University (OSU) led by Prof. Katherine Flores (now at Washington University St. Louis) (Zr-based BMGMC) [477–479], Dr. Brian Welk (OSU) along with Dr. Mark Gibson (CSIRO,

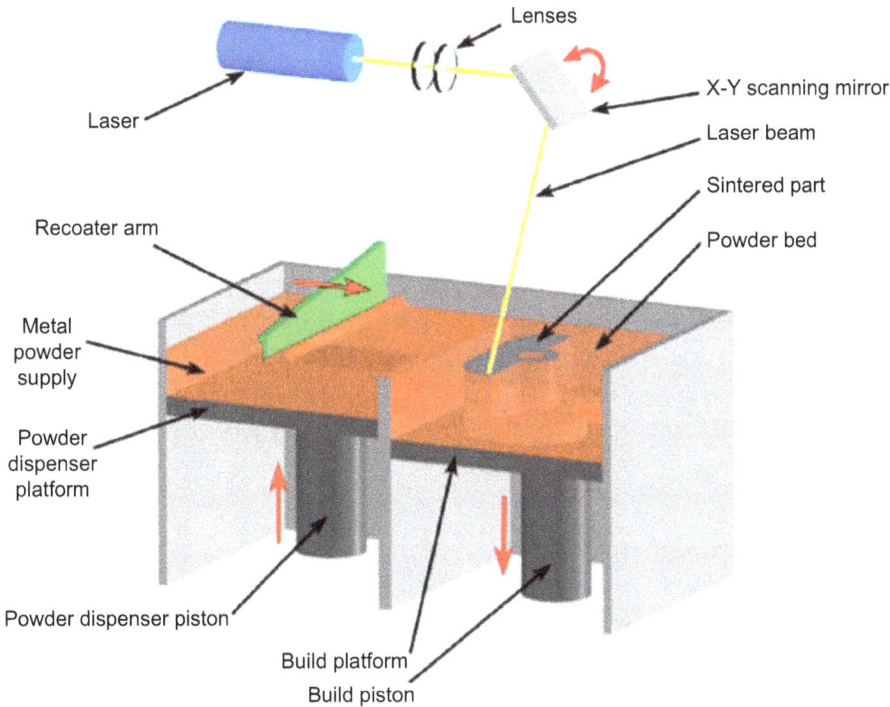

**Figure 19:** Schematic of selective laser melting (SLM) systems [83].

Australia) (compositionally gradient alloys (BMGMC–HEAs)) [481, 485], Prof. Tim Scrcombe at University of Western Australia, Perth (Al-based BMGMCs) [467–469], Harooni, A and colleagues at University of Waterloo, Canada (Zr-based alloys) [84], Prof. Scott M. Thompson at Mississippi State University (MSU) [482, 483], Prof. Jurgen Eckert at IFW Dresden (Zr-, Al- and Fe-based BMGMCs) [59, 306, 307], H-S Wang and colleagues in Taiwan (Zr-based BMGMCs) [412], Yue et al. (Zr-based BMGMCs) (Hong Kong) [87, 88] and Prof. Huang at Northwest Polytechnic University, China (Zr-based BMGMCs) [69, 70].

Advantages associated with the use of AM techniques in the processing of BMGMC are in addition to above-mentioned advantages of process as a whole. For example, in AM of BMGMC, extremely high cooling rate inherently present in the process gets exploited in a sense that it helps in by passing the nose of TTT diagrams of composites, thus enabling formation of glassy structure without "phase separation (LLT)" [92] or "NG" of crystalline phases [59]. The same high cooling rate and glassy phase formation is responsible for imparting very high strength in these systems. Similarly, LBL formation helps in the development of glassy structure in top layer and as the laser travels the path dictated by geometry of part to be produced, the preceding layer get heated to a temperature somewhere inside the nose of TTT diagram. This causes devitrification and, if controlled properly in a narrow window of time and composition, gives rise to tough and strong structure in one-step fabrication. The same process is limitation as well, as if composite (i.e., layer beneath fusion layer) spends too much time inside nose of TTT diagram, it may cause complete (100%) crystallization forming crystalline alloy which is neither desirable nor is the source of enhanced properties of laser-processed BMGMCs. Latest trend is to use compositionally gradient alloy compositions [486, 487] along with advantages of AM to fabricate a series of alloys which have a spectrum of properties as a function of composition and laser parameters (power, scan depth, scan rate, and spot size). The final composition which is best and suitable for a particular application could be chosen thereafter depending on type of application. This gives rise to not only production of BMGMC but incorporate HEAs [488, 489] in this research as well. Very recent publications by Dr. Mark Gibson [481, 485] are notable in this regard. Laser deposition has also been used to identify GFA (an age-old disputed property of glassy systems) [490].

### 2.9.13.1 Equilibrium phase diagrams

BMG traditionally were explained by the help of equilibrium diagrams, mainly binary in nature [144]. As in early days (before the establishment of three laws in last decade of last century) only binary compositions were rapidly quenched to produce glassy structure mainly for laboratory purposes and no efforts were made to understand the exact, complex and varied behavior of these alloys systems out of laboratory conditions on industrial scale. It was only after 2007 (establishment of three rules) [4, 118] that ternary diagrams were more often used to predict and understand the behavior in particular GFA of this class of materials. Most of those diagrams were based on eutectic compositions despite the fact the GFA is most often off-eutectic and is merely an empirical parameter (as discussed earlier in Section 2.4.1). These diagrams are helpful to certain extent only in understanding the behavior of BMGMCs. Since behavior of BMGMCs in actual conditions is not at all equilibrium cooling due to rapid quench, more effective way to understand and explain this is by means of nonequilibrium or TTT diagrams (explained in the next section).

### 2.9.13.2 Nonequilibrium phase diagrams (e.g., TTT diagrams)

Since most of BMGMCs do not form under slow cooling and in particular very high cooling rates are met in AM, their behavior and development of phases could best be explained by nonequilibrium, for example, TTT diagrams (Figure 15). These diagrams are very similar to TTT diagrams used extensively in steel metallurgy. The only major difference being their nose is shifted to extreme right because of their multicomponent nature [60]. Like other alloys, their behavior is dictated by alloying elements present in them. Most of the alloying elements shift the diagram to right making it easy for the glass to form. These are also helpful in calculating GFA [164]. They are also very helpful in explaining and designing of procedures for TPF of these alloys [59].

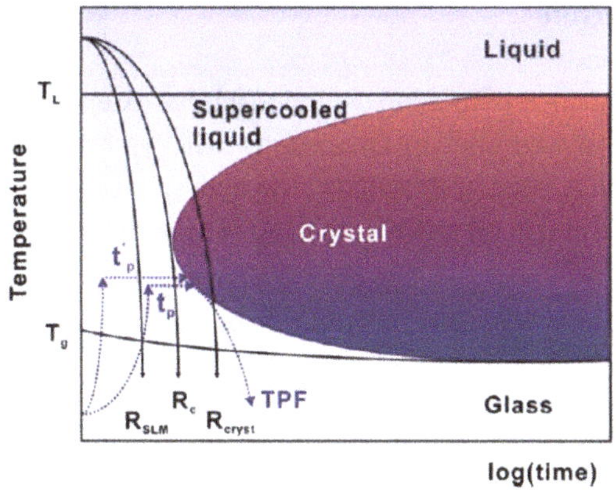

**Figure 20:** Schematic TTT diagram for BMGMCs (indicating regions (cooling rates) of monolithic glass formation as well as thermoplastic forming (TPF))[59].

### 2.9.13.3 Invariant temperatures ($T_L$, $T_m$, $T_g$, $T_x$)

Referring back to Figure 20, distinct regions on TTT diagram can be identified. This include invariant temperatures marked by $T_L$ (liquidus temperature (upper line)), $T_m$ (melting temperature (not shown and is usually equal to liquidus temperature for most alloys)), $T_s$ (solidus temperature (not shown and is usually visible on equilibrium phase diagram (temperature at which a phase transformation from liquid to solid is completed)), $T_g$ (glass transition temperature (lower line)) and $T_x$ (crystallization temperature (onset of crystallization (not shown)). There is additional information available and can be used from TTT diagrams. This includes cooling rates data. For example, the figure shows $R_c$ (which is critical cooling rate exactly necessary to form fully glassy (monolithic) structure), $R_{cryst}$ (cooling rate for crystallization). This is the rate at which if an alloy is cooled, it results in crossing the nose of TTT diagram resulting in the formation of partial or full crystals in glassy matrix (not desirable)) $R_{SLM}$ (this is critical cooling usually desirable, and is achieved, in SLM. It clearly shows that no crystallization will happen and fully glassy structure will form).

## 2.10 Model alloys

Two types of alloys namely $Zr_{47}Cu_{45.5}Al_5Co_2$ (eutectic) and $Zr_{65}Cu_{15}Al_{10}Ni_{10}$ (hypoeutectic) are studied here as model alloys.

### 2.10.1 Common phases and microstructures

Although microstructure of alloy is a function of its composition to a large extent, this section details the most commonly observed ones in Zr-based as-cast hypoeutectic (Zr > 65%) and eutectic (Zr < 50%) systems used in this study. The alloys investigated are $Zr_{47}Cu_{45.5}Al_5Co_2$ (eutectic) and $Zr_{65}Cu_{15}Al_{10}Ni_{10}$ (hypoeutectic). Their microstructures are explained below.

### 2.10.1.1 $Zr_{65}Cu_{15}Al_{10}Ni_{10}$ System

This system primarily consists of

a. $Zr_2Cu$-type tetragonal phase formed at very high cooling rates only and

b. $Zr_2Cu$ + eutectic ($Zr_2Cu$ + $ZrCu$) type phase which is formed at intermediate (6 mm/sec) to slow (1 mm/sec) cooling rates .

An inverse relation exists between eutectic and cooling rate. The amount of eutectic increase as cooling rate is decreased:

$$\text{eutectic} \propto \frac{1}{\text{cooling rate}} \qquad (5)$$

Other phases which are present in these alloys are $\tau_3$ and $\tau_5$. However, these are not observed as there is Ni in the system replacing some of Cu. A second prominent effect which is observed in these systems is the effect of Zr content. Table 1 shows Zr content and its effect on phase development at constant withdrawal velocity of 6 mm/s.

**Table 1:** Qualitative analysis of different phases present in Zr-Cu-Al-Ni alloy system [52].

| S. no. | Zr content | Crystalline precipitates | | Glassy substrate |
|--------|-----------|---------------------------|---|------------------|
| 1 | $Zr_{57}$ | $Zr_2Cu$ type (similar to $Zr_{60}$ (tetragonal) but different in morphology) | – | In percentage |
| 2 | $Zr_{55}$ | $Zr_2Cu$ type (similar to $Zr_{60}$ (tetragonal) but different in morphology) | – | In percentage |
| 3 | $Zr_{52.8}$ | Nil | – | 100% (monolithic BMG) |
| 4 | $Zr_{50.1}$ | ZrCu type (monoclinic) | – | In percentage |

A third important observation in this class of alloys is evolution of percentage of crystalline phase, its morphology and percentage of glassy matrix with cooling rate (expressed in terms of withdrawal velocity). This is elaborately explained in Table 2.

**Table 2:** Qualitative analysis of effect of cooling rate on evolution of different phases [52].

| S. no. | Cooling rate | Crystallite (percentage) | Morphology | Glass (percentage) |
|---|---|---|---|---|
| 1 | 6 mm/s | Nil | Nil | 100% |
| 2 | 4 mm/s | $Zr_2Cu$ + ZrCu eutectic (<100%) | Spherical | <100% |
| 3 | 3 mm/s | $Zr_2Cu$ + ZrCu eutectic (<100%) | Spherical | <100% |
| 4 | 1 mm/s | 100% $Zr_2Cu$ + ZrCu eutectic | Spherical | Nil |

This also confirms the relation observed in eq. (4). In addition to that, in this class of alloys invariant temperatures have been observed to have the following behavior.

Glass transition temperature $(T_g)$ is observed to have inverse relation with Zr content (Figure 21(a)).

Note that $Zr_{55}$ is at $T_L$ = 1,157 K which is eutectic temperature. However, $Zr_{52.8}$ is best glass-forming composition which is off-eutectic. This is contradiction. However, it is empirical relation and experimental result indicate that none $\Delta T_x$, $T_{rg}$ and/or $y$ best expresses GFA in these systems. This is a typical case of the presence of best GFA at off-eutectic temperature as is witnessed by earlier observations [169]. Similar behavior is observed previously for some Cu- and La-based BMGMCs. However, more research (e.g., variation of percentage of ductile phase and its number density and its relationship with GFA) is needed to verify this hypothesis in hypoeutectic Zr-based systems. Another important fact observed in these systems is effect of variation of GFA with Nb content. Nb is observed to have very prominent effect on fluidity and mechanical properties as controlled by tuning of microstructure in these alloys [154, 491, 492]. For example, in a study conducted by Sun et al. [491] it was shown that addition of Nb up to maximum of 15 at% causes precipitation of β-Ti-like dendrite phases in glassy matrix. These dendrites are few in number at 5 at% and tend to increase with increasing Nb content with the formation of other quasicrystalline particles. Their behavior is qualitatively shown in Table 3.

This study confirms their observations in other similar efforts aimed at tuning other properties by controlling dendrite parameters (type, size and shape) and microstructure [493, 494]. It is also observed in another study by Professor Inoue and colleagues that crystallization process of Zr–Ni–Cu–Al MG is greatly influenced by adding Nb as an alloying element [154]. Based on the results of the differential scanning calorimetry (DSC) experiments for MG $Zr_{69-x}Nb_xNi_{10}Cu_{12}Al_9$ ($x$ = 0–15 at%), the crystallization process takes place through two individual stages. For ($x$ = 0), metastable hexagonal ω-Zr and a small fraction of tetragonal $Zr_2Cu$ are precipitated upon completion of the first exothermic reaction. The precipitation of a nano-QC phase is

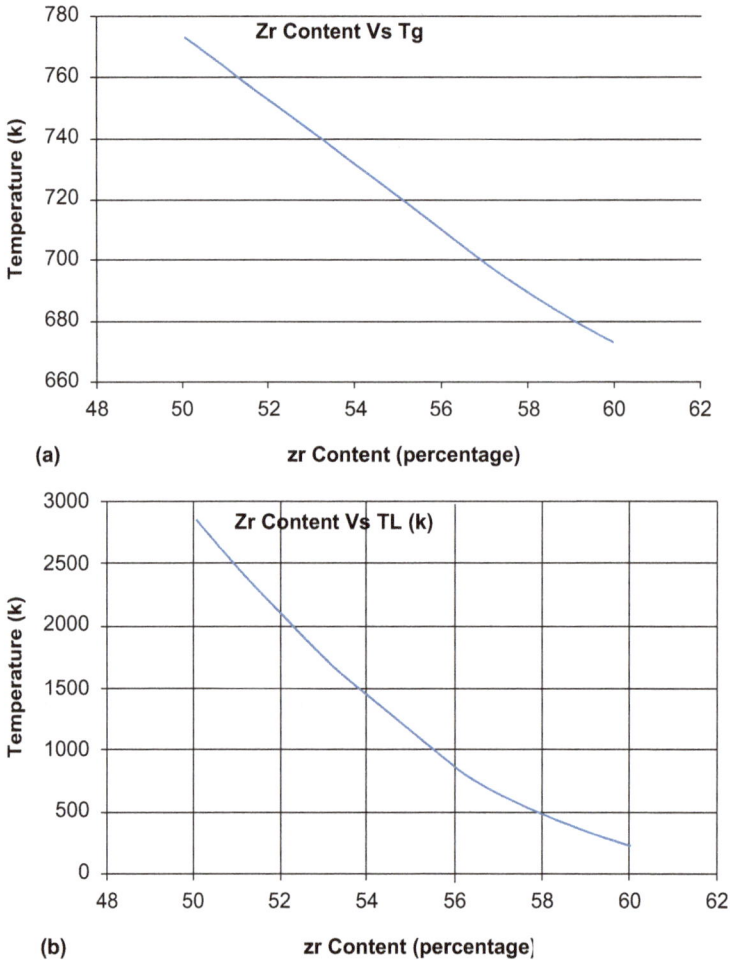

**Figure 21:** (a) Graphs showing relation between glass transition temperature ($T_g$) and Zr content and (b) graphs showing relation between liquidus temperature ($T_L$) and Zr content [52].
$T_x$ is crystallization temperature (onset of crystallization) is independent of composition.
$T_m$ is melting temperature that is constant for all alloys at 1,094 K indicating that all alloys are formed at the same constant eutectic reaction temperature.
$T_L$ is liquidus temperature, which shows nonlinear (decreasing trend) dependence on composition (Figure 21(b)).

detected when Nb content is raised to 5–10 at%. Similar trends were observed in studies conducted by Professor Eckert's group at IFW, Dresden [492, 495]. The ongoing research on this class of materials shows and tally with the observations made earlier proving grounds for the validity of hypothesis that nucleant serve as sites for copious nucleation of ductile phase dendrites [28].

**Table 3:** Qualitative analysis of effect of Nb content on evolution of different phases and ultimate fracture strength ($K_{1C}$) [52].

| S. no. | at% Nb | β-Phase dendrites | Quasicrystalline (QC) particles | Ultimate fracture strength ($K_{1C}$) (MPa) |
|--------|--------|-------------------|--------------------------------|---------------------------------------------|
| 1 | 5 | Low percentage (<100%) | Nil | 1,793 |
| 2 | 10 | Intermediate percentage (<100%) | <100% | 1,975 |
| 3 | 15 | High percentage (<100%) (fully grown 3D morphology) | <50% | 1,572 |

## 2.10.1.2 $Zr_{47}Cu_{45.5}Al_5Co_2$ system

This is the system in which not only famous ductile phase B2 bearing regular bcc structure is observed but its transformation products B19′ (bearing martensitic structure) is also observed [26]. In these ZrCu-based alloys, strain hardening rate is enhanced and plastic instability is suppressed due to a martensitic transformation of B2 to B19′. In fact, shape memory effect (which is evolution of unique property [51]) is also observed which is due to simultaneous reversible deformation of strained B19′ (present in certain percentage) along with a certain percentage of regular strain-free B2 ZrCu bcc. The presence of these two fractions causes a tuning effect which gives rise to shape memory phenomena (i.e., strain-free lattice (regular bcc) can be reversibly changed to strained lattice (martensite) by the application of heat, causing restoration of shape [496–498]). The detailed mechanism for a system studied by Wei-Hong and coworkers [26] is given below. Shape memory effect along with GFA is associated with martensitic transformation of B2 to two monocline structures:

a.  a base structure (B19′) with $P2_{/m}$ symmetry and

b.  a superstructure with $C_m$ symmetry.

Transformation temperature hysteresis of ZrCu-based shape memory alloy (SMA) is large while thermal stability is poor. Grain size is observed to have inverse relation with Co content (expressed in percentage). Average grain size of $Zr_{47}Cu_{45.5}Al_5Co_2$ is 6 μm. The microstructures observed in these alloys are $Co_2Zr_3$ and B2. Transmission electron microscopy (TEM) shows that both austenite and martensite coexist which is an indication of the fact that Co ensures the stability of martensite over a large temperature range. In other words, martensite transformation temperature becomes low. This martensite exists in $C_m$ symmetry.

**Note:** Rietveld refinement shows that

a.  At normal conditions: in intermetallic compounds (IMCs) $Zr_{50}Cu_{50}$, two types of martensite exist: B19′ and $C_m$. Both have a certain volume fraction present in conjunction with each other. B19′ have 27% $V_f$ while $C_m$ have 73% $V_f$.

b.  Under compositional contract conditions:

a. When the content of Al atom substituting for Zr atom is smaller than 9.375% (mole fraction) (Al < Zr 9.375%): Austenite phase could form a martensite base structure during quenching or straining (popularly known as stress-induced martensitic transformation/TRIP. This phenomenon is observed not only in Zr-Cu-Al-Co systems but also in many other systems [21, 22, 25, 249, 253, 254, 256, 499–502]).

b. When Al > Zr 9.375%: Austenite phase could form a superstructure ($C_m$).

c. Co-doing: Another important phenomenon is "co-doping" of Al and Co. This reduces the formation of B19′, thus makes it even more difficult to find B19′ martensite.

d. "One-step" transformation: Another notable observation is that only "one step" transformation occurs; that is, B2 transforms directly to $C_m$. Only one exception is $Zr_{47.5}Cu_{46.5}Al_5Co_1$ in which case B2 first transforms to B19′ and then B19′ transforms to $C_m$ phase upon cooling. In this case, $M_s = 309$ K while, $M_f = 275$ K.

*Addition of aluminum* causes a decrease in martensitic transformation temperature ($M_f$) until the Al content reaches a value slightly greater than 6%. However, $M_s$ remained almost constant. *Addition of cobalt (Co)* $M_s$ temperature rapidly decreases with addition of Co content. When the addition of Co increases to 2%, the martensitic transformation temperature ($M_s$) and transformation hysteresis change invariably. This happens as a result of variation of intrinsic factors, that is:

a. Increase in unit cell volume and

b. Decrease in electron concentration with increasing Co content (because Co has small atomic radius and high electron concentration)

Mechanical properties: stress–strain curve of $Zr_{47.5}Cu_{45.5}Al_5Co_2$ is shown in Figure 22.

*Compressive strength* of alloy increases with increase of Co content. This is attributed to shear induced martensitic transformation from cubic B2 to a monoclinic martensite phase ($C_m$) which imparts an appreciable work hardening capability. *Fracture strain* increased from 0.73% to 1.76% as Co content varies from 0.5% to 2%. Fracture surface analysis revealed that at lower concentrations, intergranular fracture dominates. As cobalt content increases to 2%, ductile fracture features started to appear. Surface at this concentration is characterized by a lot of faults and tearing ridges which is indicative that plastic deformation happens first prior to failure. The addition of Al and Co significantly refines grains. The martensite plates become fine. The substructure of alloy is mainly (001) compound twin and martensitic variants are (021) type-1 twin related. In a microstructure of fractured surface observed under scanning microscope, charged surface indicates fracture features (Figure 23). (Note: from crystallographic view point, B2 is cubic in nature while B19′ in its both morphologies (i.e., $P_{m/2}$ and $C_m$) is monoclinic.)

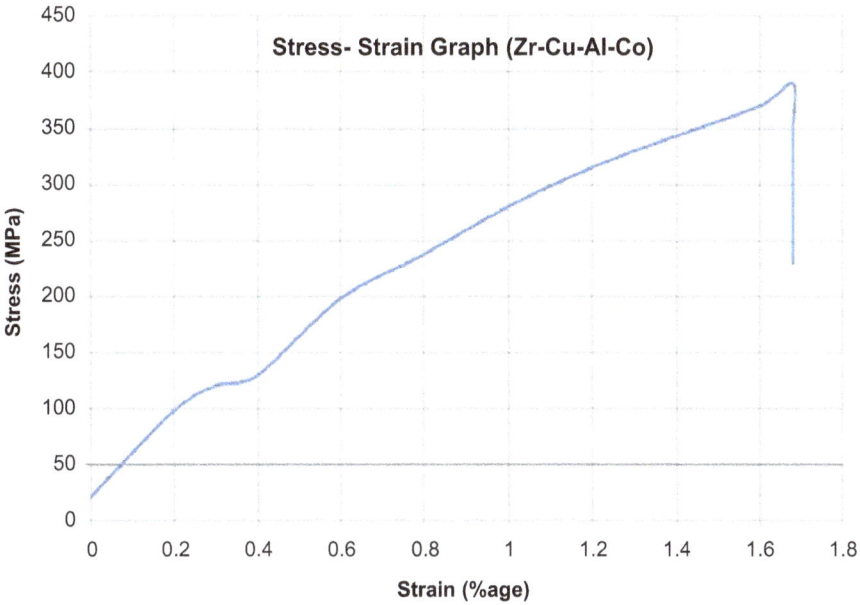

**Figure 22:** Stress–strain graph of $Zr_{47.5}Cu_{45.5}Al_5Co_2$ [26].

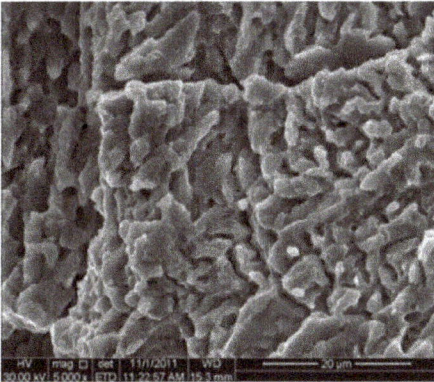

**Figure 23:** SEM image of fracture surface of $Zr_{47.5}Cu_{45.5}Al_5Co_2$ [26].

## 2.10.2 Mechanical properties

Like microstructure, mechanical properties of BMGMC are strong functions of composition. Due to the alloy systems under investigation, a contrast, as observed in their varied composition is described here only. For example, in $Zr_{47.5}Cu_{45.5}Al_5Co_2$, Table 4 shows the 0.2% offset yield stress ($\sigma_{0.2}$ (MPa)), maximum stress (UTS) ($\sigma_b$ (MPa)) and fracture strain ($\delta$/%) of different compositions of aforementioned alloy.

**Table 4:** Mechanical properties of different ZrCu-based eutectic systems [26].

| S. no. | Alloy | Yield stress ($\sigma_{0.2}$ (MPa)) | Maximum stress ($\sigma_b$ (MPa)) | Fracture strain ($\delta$/%) |
|---|---|---|---|---|
| 1 | $Zr_{48}Cu_{47.5}Al_4Co_{0.5}$ | 136.25 | 181.08 | 0.73 |
| 2 | $Zr_{47.5}Cu_{46.5}Al_5Co_1$ | 275.84 | 311.82 | 0.75 |
| 3 | $Zr_{47.5}Cu_{45.5}Al_5Co_2$ | 367.95 | 392.59 | 1.76 |

Similarly, Figure 24 shows the compressive stress–strain curves of different compositions of ZrCuAlNi alloy with and without Nb content at room temperature (maximum till 15 at%).

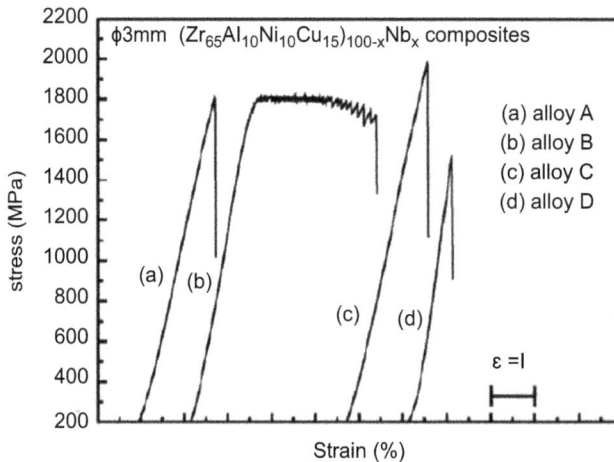

**Figure 24:** Room-temperature compressive stress–strain curves of as-cast $Zr_{65}Cu_{15}Ni_{10}Al_{10}$ with different percentage of Nb (alloy A (Nb = 0 at%), alloy B (Nb = 5 at%), alloy B (Nb = 10 at%), and alloy C (Nb = 15 at%)) [491].

It shows a dramatic behavior of change in yield stress, maximum stress and fracture stress for each composition. Alloy A with zero percentage Nb has a good yield stress coinciding with maximum stress. Alloy B with 5 at% Nb content shows serration behavior (plastic deformation) of continuous drop and gain in stress after yield stress which continues till a certain strain value before decrease in stress and failure. Alloy C (10 at% Nb) shows an appreciable increase in yield and maximum stress values but the fracture behavior is similar to alloy A without any serration and finally in the end, alloy D (with maximum 15 at% Nb) shows a dramatic decreases in ability to absorb stress before and after failure as compared to all other values. This is attributed to the development of certain IMCs and other constituents at higher alloying element content which might have caused decrease in stress.

## 2.11 Very recent trends, approaches and triumphs

Some of the modern approaches to the problem of achieving ductility and toughness are fundamental in nature based on basic understanding and comprehension of engineering and metallurgy. For example, a recent study details the size effects on stability of shear band development and propagation. This interesting review documents very recent developments and progresses in ductile BMGMC in the form of important phenomena of shear banding which ultimately results in increased ductility and toughness in otherwise brittle solids [503]. As discussed in Section 2.11, formation of stress-induced transformation (TRIP) inside a ductile phase dendrite is another promising way of achieving large ductility while maintaining high strength and hardness. Although it is relatively old idea, which was exploited some years ago by means of indentation and conventional deformations [213, 500, 504–506], it has attracted the attention of researchers as new methods of forming and transformation (especially since in situ liquid–solid transformation [28]) have evolved with time. The quest for obtaining a ductile BMG composite with enhanced optimal ductility with large enough size still continue to push boundaries of what could be achieved. In this regard, very recently, researchers at Yale University and IFW, Dresden, Germany, have made further promising progress the details of which could be found in reference [204].

### 2.11.1 Understanding structure

To understand structure of MG matrix composites, it is necessary to understand and apply theories of NG in condensed systems [507, 508]. These help to understand the basic mechanism how structure evolves in metallic glass systems. Fundamentally, it is not a homogeneous process. Small domains of $\beta$ appear at distant points in $\alpha$ and then grow at the expense of $\alpha$. Kinetics of phenomena can be described by two constants, the rate of nucleation $R^\star$ (number of $\beta$ domain appearing per unit time and per unit volume of the untransformed phase $\alpha$) and the rate of growth after nucleation, $u$ (cm/s). It is the aim of nucleation theory to calculate nucleation rate $R^\star$.

#### 2.11.1.1 Nucleation and growth
One of first attempts to understand NG in BMG was to understand stability of undercooled liquids and nature of glass transition [509]. It is reasonably believed that glass formation occurs only if upon cooling of melt below glass transition temperature,

crystal NG are avoided. Until recently, it was only possible for simple single component alloys rapidly cooled. Many studies followed which made use of fundamental theories to explain GFA [510–514], metastability [509], role of alloying elements [512, 515–517] and cooling temperature [510–512, 514, 517–521]. Recently, multicomponent alloys with deep eutectics have been developed which allow formation of MG even during slow cooling. As such, the relevant thermodynamic properties of the metastable glassy and undercooled liquid states can be directly measured below and above the glass transition temperature, respectively. The obtained data gives new insight into the fundamental aspects regarding the stability of undercooled liquids and the nature of the glass transition in these BMG.

### 2.11.1.2 Devitrification
Devitrification is a phenomenon during which a fully glassy material undergoes full or partial crystallization [522]. This may be achieved by heating or deformation or a combination. In former, material is heated to a temperature near or above glass transition ($T_g$) temperature but below crystallization temperature (thermal devitrification) while in later it is deformed at or above room temperature (deformation-induced devitrification [523, 524]) causing crystallization. This may be due to or accompanied with certain phenomena prior crystallization such as liquid–liquid transition (LLT), phase separation but final product is fully crystalline structure. Crystallization can happen by homogeneous or heterogeneous nucleation. Devitrification may be controlled and used to tailor a specific microstructure(s) in an alloy [229, 525, 526]. This can also be used to achieve certain specific mechanical properties such as size and constituent dependency [526], self-toughening [523], surface roughness [527], modification of surface to achieve enhanced fatigue strength [340, 528–532], corrosion properties [533] and thermal properties. Devitrification plays a very important part in AM [525] of BMG as well as it controls the amount, size and distribution of percent vitrified and devitrified structure in each layer which help in determining deposition strategies for manufacturing of large-scale parts by AM [534–536].

### 2.11.1.3 Polymorphic crystallization
Recently, an old concept of crystallization in MG [537] is revisited and useful results are obtained. It was earlier ascertained and determined that MG are formed as a result of suppression of diffusion [538, 539] and retardation of crystallization. Same phenomena are held responsible for polymorphism [540, 541], appearance of crystals (polymorphic crystallization [542]) in early melts and thin films [543] as a result of diffusion-assisted [537] NG. However, very recent studies [544–546] have provided more details into exact mechanisms happening. It stresses on the use of ultrafast calorimetry [542, 547] and phase diagrams to study crystallization, competing crystalline

states [548] and phases [549] and minute microscale phenomena happening. This provides a lot of detail and insight into mechanisms of crystallization in glassy melts.

### 2.11.1.4 Relaxation

Relaxation is natural physical aging of MG [117, 550–555] whereby MG undergo a structural transition. It is an old concept [556] which is readdressed and revived. This may be accompanied by or augmented with primary [557–559], secondary [557, 559–563], structural [179, 554, 564–578], thermal [180], dynamical [579], mechanical or chemical independent [580], composition dependent [581], dynamic mechanical [181, 579, 582–587], enthalpy [588], α [561, 589–592], β [589, 593–614], γ [615, 616] and Johari–Goldstein [602, 617, 618] relaxation. Each has its own characteristics and may be studied by mechanical spectroscopy [561, 583, 585, 590, 619], molecular dynamics [564, 620], quenching [621], calorimetry [601], dilatometry and density measurement [622]. Mostly beta relaxations are observed in La-/Ce-based MG [578, 589, 593, 600, 603, 604, 611, 623, 624]. These may be observed during annealing [559, 625, 626], cryogenic treatment [621], ion irradiation [268, 627–633], compositional adjustments and deformation [579, 634, 635]. Its main forms are described as under.

#### 2.11.1.4.1 Dynamical mechanical relaxation

Dynamic relaxation is an intrinsic and universal feature of glasses and enables fluctuations and dissipations to occur. These fluctuations become the reason for the onset of various behaviors and evolutions in glassy systems. This is heavily affected by alloying [579, 583–585, 587, 636], annealing [559], deformation and combination of any to apply two or more techniques together in tandem to achieve optimum properties.

#### 2.11.1.4.2 Relaxation modes

This relaxation may be dependent on composition [581, 637] and occur in different modes [559] such as one-step relaxation, two-step relaxation [626], dynamic [579], dynamic mechanical [584] relaxation (as described previously in Section 2.11.1.4.1), relaxation under ion irradiation [632] and relaxation at ultralow temperature [572, 638]. All these independently or collectively synergistically contribute effectively toward relaxation in BMG.

### 2.11.1.5 Rejuvenation–strain hardening

Rejuvenation is atomic adjustment in such a way that it promotes disorder and increases the entropy of system. It brings the system to higher energy state and is of interest in improving plasticity of material. It results in increasing the hardness of material and excess enthalpy to enthalpy of melting. However, this higher degree of rejuvenation is observed only at microscale. This extreme rejuvenation gives a state equivalent to obtainable by quenching of liquid at $10^{10}$ K/s. In most cases, it is opposite to relaxation accompanied with plastic deformation. Deformation broadens the range of interatomic distances in MG, a characteristic of disordering opposite to effects of relaxation [639]. This can exist in different manifestations such as structural [632, 640–646], mechanical [640, 647] and thermal [648–653], and may be exhibited by high-pressure [654], enthalpy relaxation [655] and techniques such as nonaffine thermal treatment [656], thermal cycling, shock compression [657], electrostatic loading [658], thermomechanical creep, cryogenic cyclic treatment [621, 624, 659–669] and heavy plastic deformation [670]. It can be measured by techniques such as ultrafast scanning calorimetry [670], heat capacity, thermal cycling, thermal history, aging and tensile testing (tensile ductility, strain hardening). It is useful to prevent fracture [671], predict mechanical behavior, irradiation behavior, and optimize processes such as TRC, cryogenic treatment and thermal cycling.

### 2.11.1.6 Ordering

Ordering is structural adjustment in MG in such a way it promotes homogeneity and decreases randomness and entropy [572, 672–674]. It may be manifested as structural order [675–681], short range order [120, 121, 682–684], short to medium range [123, 685–687], medium range order [688, 689] long range order [690, 691], icosahedral order [692, 693], and fractal packing [694]. It may be accompanied with other structural phenomena such as phase separation and LLT which are explained as follows.

### 2.11.1.6.1 Phase separation

Phase separation in MG [37, 257, 264, 513, 695] is a phenomenon which may happen prior to ordering and crystallization in some MG compositions. It results in distinctive separation of liquid into two phases such that either one, or both, crystallize independently or concurrently forming final microstructure of crystallized products [696–701]. It may be explained by classical hard sphere model [696]. It was also thought previously that phase separation only occurs in liquid [695]. However, very recent synchrotron studies have revealed its in-depth nature and provided pathways to further investigate phenomena happening alongside, prior or after it such as solid state phase separation [37, 258, 259, 348, 349, 702], LLT (explained in the next section). It may be mediated by alloying [258, 702], may assist in synthesis [262], promotes and explains heterogeneity [347], plasticity [265, 347] and GFA [703].

### 2.11.1.6.2 Liquid–liquid transition

LLT is another interesting phenomenon recently revealed by in-depth atomistic simulation and experimental (synchrotron) studies aimed at understanding nucleation, growth and crystallization in MG above liquidus line [91–93, 704–707]. It is well known that there is no phase line above liquidus in solidifying alloys and this is satisfactory to the extent that no phase or phase formation in expected. However, LLT is distinctive from the point of view that actual phase transitions exist, occur and observed in many liquids prior their solidification below $T_L$. Mostly it is temperature induced and, in some cases, it is governed by pressure. This is well documented in very recent investigations cited above. Interested reader is referred to these.

## 2.11.2 Shear bands

Shear bands [209, 214, 230, 708–711], their occurrence, formation, NG [233, 712, 713], propagation [503, 714, 715] and mechanisms by which they cause structural transformations (heating, cooling, pressure effects) is another area of great scientific interest. These are categorized as major mechanisms of deformation and failure in MG as opposed to dislocations in crystalline materials. They are also considered to be main reason of low toughness and brittleness of these materials and can be engineered and modulated to increase the same. Some of these and other effects are explained earlier. This section details very recent developments and advancements in this field.

### 2.11.2.1 Dynamics

Shar bands, their formation, nucleation, growth, propagation, arrest, creep and aging are a subject of utmost interest in recent times and are collectively studied as shear band dynamics. These minuscule features (shear bands) are the main reason of strength at room temperature, dictates their toughness by various mechanisms and are reason of their catastrophic failure as well [716]. These are governed by various mechanisms and dictated by various factors such as temperature [717], pressure and strain rate. These may be measured by high-speed optical and infrared diagnostics [718]. One of the most important concepts in this field is shear affected zone and STZ. Size, shape, nature, thickness, proximity and locations of STZ are strong factors in determining the strength and behavior of MG and their composites. Others are their multiplication [208, 209], role in fracture [235, 532, 719–721], fatigue [530, 532, 720] and plasticity [722–726].

### 2.11.2.2 Serrations

Serration is a phenomenon extensively observed during plastic deformation of MG [727, 728] and is a strong function of inhomogeneous deformation, cavitation, local heating (thermal softening and β relaxation) and shear banding. Various factors affect serrations, their occurrence, formation, role in shear band propagation [729], dynamics [730], evolution and observance. They may be studied by various techniques such as nanoindentation [731] and may be controlled or tailored by various mechanisms [732]. This is a subject of interest in present research and is actively studied. Light is also shed on the phenomena after shear banding [733] such as multiplication and avalanches [734, 735]. Serrations are affected by various factors some of most important are, effect of force, number (quantity) of force(s), amount of force and direction of force (anisotropy). These (serrations) become precursor and onset of avalanches as described previously.

### 2.11.2.3 Chemical effects

Shear bands are also heavily affected by chemical affects and compositions. These may exist in the form of interface, external confinement (including inclusions and external stimuli), diffusion and diffusion assisted interaction and internal homogeneity. The role of interface is ascertained by the fact that existing conditions at the interface of a phase govern characteristics of shear band [715]. Similarly, interface constraints at shear band which evolve their pattern and behavior as studied by micro indentation play an important part in determining, ascertaining and exhibiting their pattern and properties [328]. The role of interface also becomes important and is prominently depicted on modulating heterogeneity and plasticity in BMG [736].

### 2.11.3 High-entropy metallic glass

High-entropy MG are a special type of MG [737] in which elements are mixed in such a way that near-equiatomic ratios are maintained and a high entropy effect is produced [738, 739]. Their (including pseudo-high entropy BMG [740, 741]) formation, crystallization and kinetics are an active area of investigation [739, 740]. Glass formation in these alloys may result by design [324] and result in large GFA [742]. Some of the important properties researched for these materials are, room temperature nanoindentation creep behavior [743], dynamic mechanical relaxation [587], nanoscratching [744] and oxidation resistance [745]. Alloying elements added in these are Cr in Fe-Co-Ni-P-B [746], Fe in FeCoCrNiMn [747], Y in Gd-Co-Al-Ho [748] and Ce medium entropy BMG [749]. Two of their very important classes are pseudo-quinary $Ti_{20}Zr_{20}Hf_{20}Be_{20}Cu_{20-x}Ni_x$ [741] and quinary Ti-Zr-Hf-Be-Cu [750] high-entropy BMG. These may be manufactured by thermoplastic micro formability [751].

### 2.11.4 Nanoglasses

Nanoglasses are new kind of nanocrystalline materials [752, 753] in which mechanical properties are strong function of grain size [754], interface and interphase [755, 756], microstructural features and length scale [757, 758]. Interphase plasticity governs strain delocalization [756, 759]. Interphase also act as reinforcement and cases strengthening [760], it also controls Curie temperature if its composition is controlled by varying volume fraction of phases, nanolaminates are also formed and observed to enhance strength [761, 762] and sample size effects play an important role in strength and deformation mechanisms of nanoglasses [763]. Some of the most prominent systems studied are Cu-Zr [754], $Cu_{64}Zr_{36}$ [761], $Ni_{50}Ti_{45}Cu_5$ [764] and $Fe_{90}Sc_{10}$ [756, 765, 766].

### 2.11.5 Corrosion behavior

Corrosion behavior of MG are extensively studied recently and are heavily affected by parameters such as alloying [533, 767–775], surface finish [776, 777], process variables in processes like laser AM [87, 88, 778–783], cladding [781, 782] and magnetron sputtering [447] for thin films [432, 445, 447, 453, 465, 533, 773, 784–789]. All types of corrosion are studied with a special focus on pitting [453, 790–793], tribocorrosion [794–796], stress corrosion [797–800], corrosion fatigue [340, 799, 801] and biomedical corrosion [445, 770, 774, 783, 784, 794, 802–819] (in vivo [813, 817] and in vitro [809, 813, 817, 820, 821] in simulated body fluid [798, 818] or unsimulated body fluid conditions).

### 2.11.6 Shape memory BMG

Shape memory BMG [14, 496] are a special type of BMG in which reversible martensitic transformation of B2 to B19′ [26, 822, 823] is used as a measure to develop and control shape memory effect. Various manifestations of these alloys are used in practical applications. These and their processing such as by spray forming, rapid quenching, flash annealing, laser marking–assisted wire arc–based AM and characterizations such as by TEM [498], temperature resistance sensor arrays [701] may be manipulated, tailored and controlled by various factors such as monoclinic phase transformation [824], alloying [26, 51, 822, 825], metastability [826, 827], anisotropy [823], deformation and twinning [317, 497], cooling rate [828] and transient nucleation [829].

### 2.11.7 Limitations/research gap – new findings

Despite advances and triumphs, still there are number of unresolved issues from processing (chemistry, physics, metallurgy and engineering (tooling, machinery, etc.)), structural (phase identification and their behavior), properties (mechanical, physical and functional) viewpoint which limits their application and further use in more advanced applications, commercialization and large scale production. For example, despite being able to be produced in bulk form, still the largest ingot casted on BMGMC is just 80 mm in diameter and 85 mm in length [45]. Liquidmetal Technologies have been able to produce various types of shapes in "cast" form but these are by adopting very expensive tooling and are very thin in their profiles [60]. There are very few successful efforts to make parts with tensile strength greater than 980 MPa in Al-based BMGMCs [830]. Despite its advantages, TRC remains a novice technique for fabrication of BMGMCs of all types. Only Ti-based BMGMCs could be produced with ease because of their increased fluidity. Zr-based BMGMCs still have the biggest limitation for large-scale production as these are viscous and their transformations are sluggish because of suppressed kinetics. There is very little effort on the functional use of BMGMCs [831]. Reproducibility of these composites is another outstanding debate and contradictions exist about their behavior from laboratory to laboratory. The effect of microstructural control parameters and its tuning with variety of materials and physical parameters is not known. Lastly, AM [83, 832], though a promising technique and presently being named as "Future," has serious drawbacks (microstructure, modeling, metallurgy, mechanical properties, anisotropy, etc.) for the use of Al- [63], Fe- [57, 71, 307, 470], Ti- [474, 833] and Zr-based [69, 70, 87, 475, 478, 479] BMGMCs.

### 2.11.8 Present research – advancement since the first edition

In present research, an effort has been made to microstructurally control and tune the properties of Zr-based BMGMCs by controlling the number density ($d_c$) of ductile second phase (B2), its grain size and dispersion in bulk alloy by conventional and AM routes. This novel idea stems from the fact the inoculation of otherwise passive melt can cause precipitation of certain phases prior to other microstructures in an alloy. This can effectively be used for evolution of preferred phases thus affecting properties. It is envisaged that careful selection of potent inoculants which can best serve as sites for preferential nucleation of ductile phase only can best be used to increase their number density, and dispersion in bulk of alloy. It has been previously reported that a 3D arrangement of network of ductile phase equiaxed dendrites in bulk alloy can effectively serve as source of impediment of shear band motion and can best serve as a junction for their multiplication [12, 297]. Further, there are methods by which only high-potency inoculants whose crystal structure matches the crystal structure of precipitating phase can be preferentially selected as compared to other inoculants. This is known as "edge-to-edge matching" [834–837]. Selection of nuclei by this method and then controlled inoculation by them can serve as an effective means for increasing the number density, size and distribution of ductile phase dendrites in bulk. This fact is successfully exploited in present research. During the course of study, computational model based on probabilistic cellular automaton are proposed to be developed which will be used to predict the size, shape and morphology of dendrites and their evolution. The model takes into account the effect of crystallographic orientation and motion of liquid–solid front as well. This is proposed to be coupled with a transient heat transfer model in the melt pool of additive manufactured part (laser materials interaction region). A code of model is aimed to be developed in MatLab Simulink® and its coupling is proposed to be done by SolidWorks® and Ansys®. The results predicted by computational studies are proposed to be verified by their observation in actual fabricated samples in SLM machine. This experimental verification was proposed to be done by optical and electron microscopic analysis. Progress has been made in these studies since the first edition such as ore modeling and simulations have been performed, propositions are made, and experimental studies encompassing double and multipass laser AM have been carried out. Some new alloys and their compositions have also been proposed. An interesting area of ion beam irradiation characterization aimed at using them as plasma facing materials has been started. Details of these processes and advancements are described in Section 2.5.

# Section 3
# Additive manufacturing (AM)

## 3.1 Why additive manufacturing (AM)?

Additive manufacturing (AM) of metal parts has recently evolved as the most important, versatile and most powerful technique bearing a lot of futuristic potential of full scalability to develop complete parts in one run without any restriction of size, shape, material(s) and complexity. In addition to that, due to its inherent nature (laser [838] or high-energy (electron) beam [67]-assisted sintering selective laser sintering (SLS) or melting selective laser melting (SLM), layer-by-layer (LBL) formation and other associated processes (denudation zones, Marangoni convection and plasma)), it eliminates the need of any post treatment (heat treatment, trimming, grinding, finishing and polishing). In most cases, it is direct production of actual (to be serviced) part in full, just in one step. Minor trimming, shaking or air blasting may be needed which does not comprise the bulk process. These unique features have raised its importance and attracted a lot of attention from scientific, technical and engineering community to further promote its use, work toward its progress and remove bottlenecks in its operation.

## 3.2 A brief history of additive manufacturing

AM has been in use since long in various forms which primarily consist of shaping of metals, alloys (Fe based [470, 839] including steels [840], Al [64, 841], Mg [842], Ti [843–845] and Cu [846] based), ceramics [847, 848], glasses [849, 850] and polymers [851–855] into different forms by the use of highly focused, localized and concentrated source of energy. Traditionally, after the discovery of electricity, for metallic parts, this energy has been supplied in the form of electric arc which melts the base metal and thus provides a small consistent melt pool which is maintained by the help of external addition of material(s) (powder, fluxed wire (filler), etc.) and serves as the source of joining, deposition (coating, facing and cladding), or fabrication (repair). Modern-day AM evolved from its predecessor which was primarily laser-assisted coating of materials (hard facing, thermal barrier coating (TBC) and biomaterials coating) on other materials. It evolved into "cladding" [84–90, 780, 856–861], which was a form of welding involving continuous or semicontinuous deposition of one material on top of other to make a thick layer which serve as protective (corrosion, wear, heat, etc.) or face layer. The latest predecessor of AM is laser (rather than electric arc)-assisted "gas metal arc welding" (GMAW) or "metal insert gas" (MIG) welding [76–81] using various types (fiber [76], Nd:YAG [77, 80, 412, 862, 863]) and operating modes (continuous and pulsed [69, 864] (micro, pico [865], and femtosecond [69, 866–871])

https://doi.org/10.1515/9783110747232-003

of lasers. This setup is very close to modern-day AM setup. In AM, the only difference is being that a part is produced in a layer-by-layer (LBL) fashion from a sliced CAD pattern fed to a computer controlling the setup at the back end. The technique matured with time to make full-scale components from various types of powder feedstocks and base metals as it attracted the attention of various groups around the globe [63, 69, 70, 467, 478, 481, 872–874].

## 3.3 Principles of additive manufacturing

Although present in, and known by, different variants (direct laser fabrication, direct metal deposition (DMD) [83, 875–877], laser metal deposition (LMD) [878, 879], direct laser deposition (DLD) [482, 483, 488, 880], direct form fabrication, direct digital manufacturing, rapid prototyping (RP) [476, 881], 3D printing, powder feed method (PFM) [882], wire feed method [484], selective laser sintering (SLS) [883–886], selective laser melting (SLM) [58, 63, 306, 470, 471, 478, 887–889], laser-engineering net shaping (LENS®) [890, 891]), basic principles of AM remain the same [57, 838]. It consists of energy source (laser [838] or high-energy electron beam (EB) [67, 874]), powder (serving as bed beneath, fed alongside (co-axial) or at an angle to energy source), a drawing/pattern generation system (CAD computer), slicing system (slicing software), and a mechanical ratchet-based or mechatronics-controlled system for step-by-step downward motion of powder holding table. There are other auxiliaries of the system (such as roller, hoppers, alignment systems, switching and viewing window) which in totality form the system but they are minor as compared to main parts described previously. Although developed with respect to its mechanical and mechatronics system, it still has its shortcomings which require more research and development (R&D) specially on materials properties and their final homogeneity across whole part (in case of very large parts and in thin sections), consistency of properties in very complex alloying compositions and very little or no knowledge about melt pool, its shape, size, depth, dynamics, evolution and transient behavior affecting final solidified metal/alloy amount and quality. This and other similar drawbacks are still challenges for its utilization, adoption and full commercialization at large scale. These restrict its use primarily to repair and damage mitigation rather than full-scale neat part manufacturing. The following sections briefly highlight various major processes presently in use to form final high-quality parts.

### 3.3.1 Laser processes

Laser-assisted AM or simply laser-based AM processes are set of procedures in which the source of energy for sintering or melting and subsequent solidification from powder bed or feedstock is provided by highly localized, concentrated and focused laser light. These technologies can broadly be grouped into one of seven major classes based on the mechanism in which each layer is formed: photopolymerization, extrusion, sheet lamination, beam deposition, direct write and printing, powder bed binder jet printing, and powder bed fusion (PBF). For the purpose of this study, these processes may be classified on the basis of how metal feedstock is made available, presented to and interact with laser light. This can be powder bed processes, powder feed process or wire feed process. Further, powder bed processes can be classified on the basis of how powder consolidation or fusion can take place. Thus, it forms the basis of SLS in the former case and SLM in the latter. A brief classification of laser-based metal AM processes is presented in Figure 20.

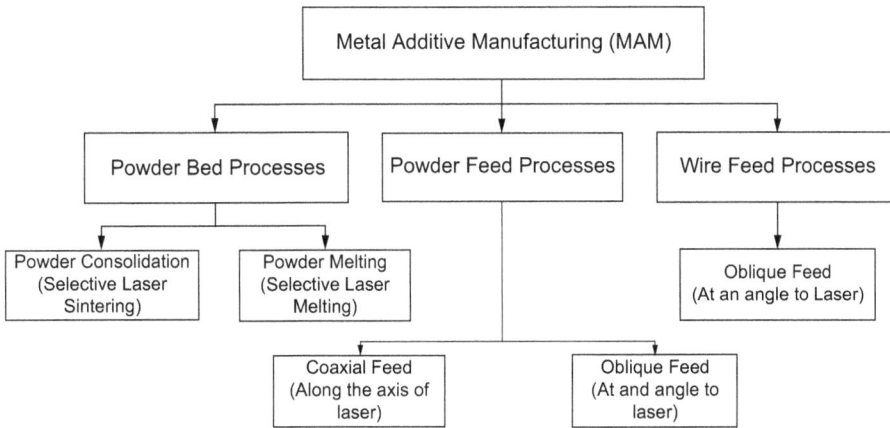

```
                    ┌─────────────────────────────────────┐
                    │ Metal Additive Manufacturing (MAM)  │
                    └─────────────────────────────────────┘
            ┌───────────────┼───────────────────────────┐
   ┌──────────────────┐ ┌─────────────────────┐ ┌─────────────────────┐
   │ Powder Bed       │ │ Powder Feed         │ │ Wire Feed           │
   │ Processes        │ │ Processes           │ │ Processes           │
   └──────────────────┘ └─────────────────────┘ └─────────────────────┘
      ┌───────┴────────┐                              │
┌──────────────┐ ┌──────────────┐          ┌─────────────────────┐
│ Powder       │ │ Powder       │          │ Oblique Feed        │
│ Consolidation│ │ Melting      │          │ (At an angle to     │
│ (Selective   │ │ (Selective   │          │ Laser)              │
│ Laser        │ │ Laser        │          └─────────────────────┘
│ Sintering)   │ │ Melting)     │
└──────────────┘ └──────────────┘
       ┌──────────────┐        ┌──────────────┐
       │ Coaxial Feed │        │ Oblique Feed │
       │ (Along the   │        │ (At and angle│
       │ axis of      │        │ to laser)    │
       │ laser)       │        └──────────────┘
       └──────────────┘
```

**Figure 25:** Classification of laser-based additive manufacturing processes.

Every variant of AM has its own significance and applications. Their behavior can be classified into two major groups of variables. (a) Production/machine variables and (b) process/material variables. The former include the type of machine, type of laser ($CO_2$, fiber, etc.), mode of operation (continuous or pulsed), mode of impingement on target, size of machine, scan depth, scan rate, scan speed and so on, while the latter may include, type of target, its shape, size, geometry, form (solid, powder, etc.), properties prior to AM, properties during AM, properties after AM and its intrinsic and extrinsic properties. Both these fundamental types of variables and their optimization are very important in order to get best properties and part quality out of AM process. Unfortunately, there is no one "rule of thumb" which can dictate the selection of a particular process best matched for an alloy. Due to noviceness and still infant nature of process very little data is available on varied types of alloy systems which can prove one material's suitability to be best with a particular type of process. There have been generalized classifications [83, 886, 892], but they all are based on empirical and irrational data and ideas. The advent and use of advanced modeling and simulation technologies [67, 472, 482, 893] to understand transient and incipient processes happening in AM have helped a lot in paving the way for the optimization of process as a whole and its suitability for a particular group of metals and alloys. However, the best optimized selection of a proper process which is best suited for an alloy or material is still at stone's throw and is evolving with time as more and more experimentation is being carried out.

### 3.3.1.1 Selective laser sintering

SLS, as the name suggests, is a laser-based AM process in which interaction of laser with powder metal bed causes localized sintering which is evolved in LBL fashion to finally fabricate a complete part which carries the impression of CAD drawing of component fed to machine via processor in a preceding step. The process was invented in 1986 at the University of Texas when Dr. Carl Deckard was awarded the first patent for his invention [894]. The process was first developed for the fabrication of 3D components (mainly prototypes for *audio-visual help* and *fit to form* test) from polymers and nylon. With the passage of time, it was extended to metals and alloys to manufacture *functional prototypes* and *rapid tooling*. The impulse in the process came from involvement of industrial clusters and computational techniques performed in laboratories to optimize the process parameters and enhance the efficiency of sintering machine [886]. Overall, aim in these processes is not to melt but heat the powder to a temperature below melting point in conjunction with other mechanisms (particle size, density, distribution, pressure, binder, etc.) which promotes their agglomeration. Based on mechanisms of how they get heated and microstructuring, SLS could be broadly subclassified into three classes [884]:

- Direct method: one component
- Indirect method: one component + polymer
- Two-component method

These are briefly described here. *Direct method: one component:* This is also called single component solid state sintering. In this, laser energy is tuned to such an extent that it causes a temperature rise close to melting point of metal power but does not exceed it. This is done to cause binding at the interfacial grain contact area. This mechanism is called "particle fusion below melting point." This process involves the formation of a neck at contacting the surface of two adjacent particles which results in reduction of the surface area, thus causing an increase in tendency of powder to aggregate. The driving force for this is reduction of free surface energy of particles and densification is proportional to this reduction. The limitation with this process is precise control of metal temperature as well as particle size. For higher temperatures or complete melting, one finds that – owing to higher viscosity and surface tension effects – the molten metal tends to form a spherical ball-type structure which has dimensions larger than particle size. On top of it, as laser spot is usually lager than particle size, many particles get melted together and a bigger spherical droplet is formed. Because of large size of this spherical droplet, it is connected with other droplets only at certain points on its contour. Therefore, it is very likely that porosity will be formed in final sintered structure. This process is workable only at slow speeds but is difficult to carry out at high speeds encountered in typical industrial setup and usually post processing to eliminate porosity is required (not a desirable feature). *Indirect method: one component + polymer*: This method involves application of thin polymer coating on powder particles prior to laser sintering. This results in reduction of "balling effect" encountered in direct processes. In this, polymer behaves like a low-temperature phase serving as binder. Overall binder content is about 1% by weight. Due to its high infrared absorption and low melting point (~ 150 °C), polymer melts in advance to metal powder (low infrared absorption and high melting point (~ 1,488 °C) and connects the metal particles together without causing melting. Final sintered parts are about 45% porous in green stage. The density of part may be further increased by its post treatment. The complete process (post treatment) comprises three stages and can be carried out in one furnace: (a) burnout, (b) high-temperature sintering and (c) infiltration. During first step, temperature is raised to 300 °C at which remaining polymer burns out followed by sintering at higher temperature (>700 °C). This causes formation of porous skeleton of metal particles. Final densification may be achieved by infiltration with copper at 1,083 °C at which Cu infiltrates the skeleton by capillary action. Final part is about 60% structural material and 40% Cu. Main **disadvantages** with the process are it is slow, time consuming and the desired density may not be achieved as a result of shrinkage. *Two-component method*: This is a typical process used with two metals: one with higher temperature ($T_2$) called structural metal and other with low temperature ($T_1$) called binder. In this process, the laser energy is tuned in such a way that it reaches a temperature ($T$) between melting points of two metals: $T_1 < T < T_2$. This causes the binder metal to melt and flow under the action of forces such as liquid pressure, viscosity and capillary action through the pores between the solid particles. Therefore, pores are reduced and strength of structure remains intact.

### Case 1: If $d_1 < d_2$

For an enhanced densification, the grain size of binding particles ($d_1$) (metal 1 with $T_1$) should be smaller than that of structural material ($d_2$). The reason for this is the difference in melting enthalpy of particles of different sizes. Bigger particles have higher melting enthalpy due to their large size, become less susceptible to melting and therefore are well wetted. On the other hand, particles of small size have lower melting enthalpy, lose their structure completely and form a cluster structure Thus, larger particles are well wetted by smaller particles.

$$\text{Particle size} \propto \text{melting enthalpy} \tag{5}$$

**Case 2: If $d_1 > d_2$**

Open porous structure will be formed. This could also happen due to lower laser energy causing an effect called "residual porosity." On the other hand, if laser power is too high, excess liquid formation may produce compact distortion (not desirable). Therefore, (a) an optimum laser power (energy) and (b) aspect ratio of binding material to structural material are very important parameters for this process. Over the course of whole process (total time for sintering), three distinct stage may be identified. *Stage 1*: During this, melting, wetting or liquid flow, rearrangement and densification happens. *Stage 2*: This is dominated by pore removal and solute participation. In this, capillary forces compress the particles together resulting in volume shrinkage. *Stage 3*: In this last stage of solution precipitation, interconnected pores pitch to form isolated pores.

Overall, first stage is dominated by rearrangement, second is by shape accommodation while filling of pores and grain coarsening happens in final third stage. In laser sintering, usually a thickness of few hundred μm/layer is achieved. The operating details of process may be found elsewhere [63, 470, 886]. It is rarely used for manufacturing of bulk metallic glass (BMG) and their composites (BMGMC) because of their multicomponent nature and binder issues.

### 3.3.1.2 Selective laser melting

This is a process which is distinctly identified by complete melting of a region of powder bed at which laser–matter interaction occurs, followed by maintaining of this liquid melt pool and its traversing all through the shape of part to be produced at all times of AM. This distinctively gives it its name, SLM. From a conception point of view, it is very much similar to SLS described in previous section. It consist of 3D drawing (usually in CAD) of part to be produced. This 3D drawing is sliced using a slicer (a software) usually in layers of 20–100 μm thick. This creates 2D image of each layer. This file is in standard stereolithography (.stl) format accepted by industry. After this, file is imported into another file preparation software package (machine integrated) which assigns parameters, values and physical supports to the file such that it could be interpreted by different industrial AM machines. However, from a mechanism point of view, it is completely different that it involves complete micro-macro-scale melting (fusion) followed by holding and then solidification. This results in solidification microstructure of metal or alloy to be deposited which follows standard metallurgical principles (phase diagrams, liquid–solid transformations, solute partitioning, crystallography, etc.) for its progression and formation. The history of SLM goes back to 1995 when Dr. Wilhelm Meiners and Dr. Konrad Wissenbach of Fraunhofer Institute ILT in Aachen, Germany, were awarded the first ILT SLM German Patent DE 19649865 [895]. Since then these researchers are the key names in further development, progression and advancement of this technology from different platforms founding numerous companies and industrial clusters. Detailed mechanical (laser metal interaction (heat generation and transfer)) and metallurgical mechanisms (alloy solidification, phase transformation, etc.) are subject part of this study and will be described in detail in later sections. This section merely concentrates onto introducing the basics of SLM process from process and operation view point.

A typical SLM process is illustrated in Figure 14. It consists of "powder hopper" or dispenser bed which carries the load of powder. An elevation mechanism below powder reservoir lifts a prescribed amount of powder above the level of build plate which is then spread in thin even layer over the build surface by a relocator mechanism (scraper, roller or soft squeegee). Another variant (as highlighted) may supply powder from above the build platen (more easy configuration and used in American machines). The thickness of powder layer is typically 10 and 100 μm. A high-energy source (laser) is made to impinge on the surface of powder bed which causes its melting due to heat generated. These lasers could be of various types (as described earlier) but normally they are fiber lasers with wavelength 1.06–1.06 μm and power on the orders of hundreds of watts. After the first layer (which is melted) is solidified as laser traverses the complete scan over whole part dimensions fed in the form of CAD file, the same elevator mechanism lifts the powder bed or hopper supply powder from top again to make a thickness of few microns, scraper or roller smooths out the powder layer and process is repeated again. This result in layer by layer fashion buildup of part conforming to dimensions and geometry fed at earlier stages (CAD level). The process is usually carried out under protective (Ar) atmosphere to avoid oxidation and contamination. There are over 50 process variables which needs to be understood and controlled to ensure the, and achieve at good quality part. The detail of all would be exhaustive. However, they can be generally classified into four categories: (1) laser and scanning parameters, (2) powder material properties, (3) powder bed properties and recoat parameters, and (4) build environment parameters. These can be further classified into controllable parameters that can be manipulated during a build process and predefined parameters that are determined at the start of a build and remain essentially set throughout the process. A detail treatment of these could be found in excellent reviews listed at the end [58, 896, 897].

### 3.3.1.3 Direct energy deposition (DED) (powder feed method (PFM) and wire feed method (WFM))

DMD, DLD, PFD or WFM is form of direct energy deposition (DED) processes in which a concentrated source of heat (laser/EB (explained in the next section)) is made to impinge on a surface which is utilized for melting along with in situ delivery of feed material (powder/wire) for LBL fabrication of complete part or single to multilayer cladding/repair. The process has evolved as major competitive method for (i) rapid prototyping of metal parts, (ii) production of complex geometries, (iii) cladding/repair of precious metallic components and (iv) manufacture/repair in recesses and difficult-to-approach regions. Like other processes, it also involves thermal/fluidic and transport phenomena which occur in different regions:

a) Laser metal interaction (heat generation)
b) Development of melt pool (fusion)
c) Shape, size, configuration and traversing of melt pool (heat and mass transfer) and
d) Metal deposition (solidification – NG and part-scale heat transfer).
e) Microstructural evolution (solute diffusion and capillary effects)

The history of process again dates back to same period of the mid-1980s and 1990s when other processes (variants of DED) emerged on the global surface. Principally, they all consist of laser metal interaction in one way or the other; that's why their fundamental principle does not change and remain the same. Machine setup in these processes is relatively simple. It consists of a built-in software which automatically checks most sensors. Powder hoppers are filled and a build substrate is positioned below them on a stationary (3-axis systems) or a rotating (5+-axis systems) stage. This increases the ability of machine to produce more complicated shapes [83]. Like PBF processes, DLD process also comprises various operating/process parameters. Setting, monitoring and controlling of these parameters are dictating pivots to ensure process and part quality to a large extent. Again, these could be categorized into two main types: (a) machine/process parameters and (b) material parameters. Most important parameters are:

- *Laser/substrate relative velocity (traverse speed):* This dictates the length of time taken by a DLD to build certain geometry. It usually is on the order of 1–20 mm/s.
- *Laser scanning pattern:* It dictates the laser position and height-wise positioning of the substrate via numerical control set by an operator.
- *Laser power:* This is the total emitted power (in watts) from the laser source. It is typically on-the-order of 100–5,000 W.
- *Laser beam diameter:* These are usually on the order of 1 mm.
- Hatch spacing:
- *Particle/powder feed rate:* This is the average mass of particles leaving the DLD nozzle per unit time (typically 1–10 g/min). This usually should be continuously monitored and nozzle cleaning and maintenance be performed if this rate is impeded.
- *Interlayer idle time:* The interlayer idle time is the finite time elapsed between successive material/energy deposits.

Generally speaking, these operating parameters are *material-dependent* and vary in conjunction with (a) *DLD machines (e.g., number of nozzles, nozzle design)* and (b) *operating environment.* **Powders** can vary in (i) size, (ii) shape and (iii) their production method. For most laser deposition processes, powders are larger in size as compared to those used in fusion processes (as there is no constraint (distance) between them and source of heat (laser)). These typically are 10–100 μm and are spherical in shape. Spherical-shaped particles have the advantage that they retard entrapment of gas in melt pool (as they are introduced in it) and thus result in final part with little or no porosity. For production of alloys, typically gas- or water-atomized or plasma-rotating electrode processed powders are used [832]. Build chamber in the case of DLD is enclosed to provide laser safety; however, it is not necessary to evacuate it and fill it completely with insert gas to provide shielding (as is done in case of SLS/SLM processes). For nonreactive metals, shield gas directed at the melt pool suffices while for reactive metals (Ti and Nb), chamber is flooded with an inert gas (Ar or $N_2$) after necessary vacuum to provide additional shielding and safety [83]. Detailed theoretical [482] and experimental [483] description of process is given in two very recent articles by Professor Thompson and his group at Mississippi State University. Depending on how raw material is fed, it may be divided into two types, namely PFM or wire feed method. There is a third variant known as LENS® which differs from both aforementioned in that the delivery of feed material could be co-axial or oblique (at an angle) to laser/EB direction (as described in Section 3.3.1.4).

### 3.3.1.3.1 Powder feed method

PFM is a variant of DED [482, 483, 898–900] in which powder is fed directly to concentrated source of energy (laser). This may be fed co-axially (direct) or at an angle (oblique) to the source of energy using hopers [899] unlike powder bed fusion [901]. Metallurgy [83], heat transfer and fluid flow [902], chemistry, thermal [903], and processing history and measurement procedure and techniques (such as in situ thermal monitoring [904]) play an important part in process operation, optimization, final part and product quality and its mechanical properties [905]. The resulting product (part or component) has certain characteristics which are unique to this process such as anisotropy or directionality of properties, more dense structure in case of co-axial feeding, homogeneity and zero or low porosity [906, 907] in applications such as biomaterials [908], materials for energy, automotive and aerospace industries. This is extensively studied by both modeling and simulation [909] and experimental approaches and results conforming to observation have been devised.

### 3.3.1.3.2 Wire feed method

Wire feed method, as the name suggests, is a process in which wire is fed into fusion zone alongside laser or at an angle to it [484, 899, 900, 911–913]. Factors such as laser energy density [914], height control [911], wire feed rate control, type, size, shape and geometry of wire play a significant role in controlling the process and dictating final part quality. A variant of process involving simultaneous powder and wire feeding has also been tried and found useful to obtain certain properties as compared to single feed method [900].

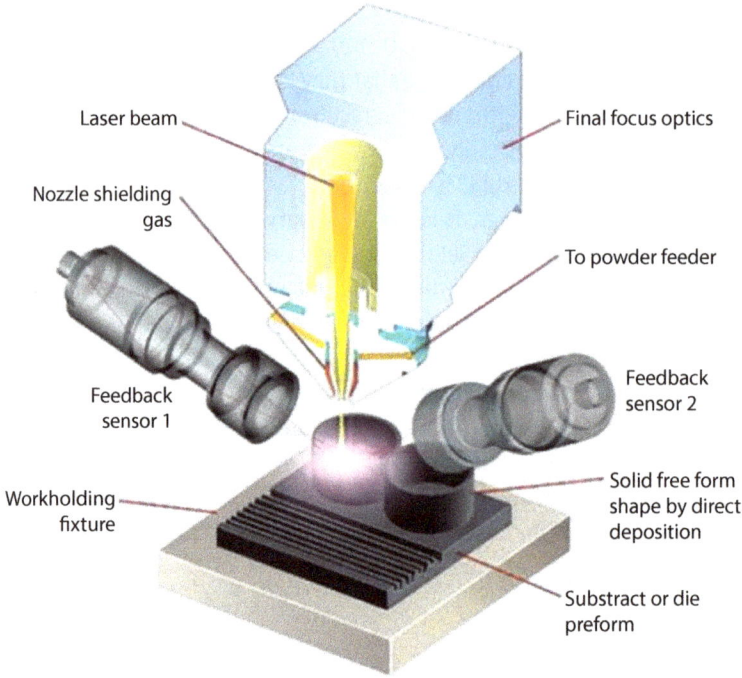

**Figure 26:** Direct energy deposition (powder feed method using laser) [910].

### 3.3.1.4 Laser engineering net shaping (LENS®)

LENS® (Figure 27) is a special variant of melting/fusion processes in which powder is preferentially fed from overhead hoppers coaxially with laser in such a way that it goes directly to melt pool and motion of "combined laser–powder feed head" keeps traversing its path dictated by CAD geometry at the back end which finally results in a three-dimensional shape [777, 890, 915–919]. The process was developed at Sandia National Laboratories in 1996 when Jeantette et. al. were awarded the first patent (US Patent 6,046,426) and derives its name from there [920]. The details about process description [917, 918], process optimization for surface finish and tuning of microstructural properties [777] are described by original inventors in articles thereafter. Owing to excellent configuration, procedure by which powder is introduced into melt pool and how it interacts with laser has evolved into new process and have found excellent applications in variant fields. These include, functionally graded materials [921] and manufacturing of parts with spectrum of properties across their cross section which can be tailored to arrive at particular property at a specific location.

**Figure 27:** Laser-engineered net shaping (LENS®) of turbine blade.

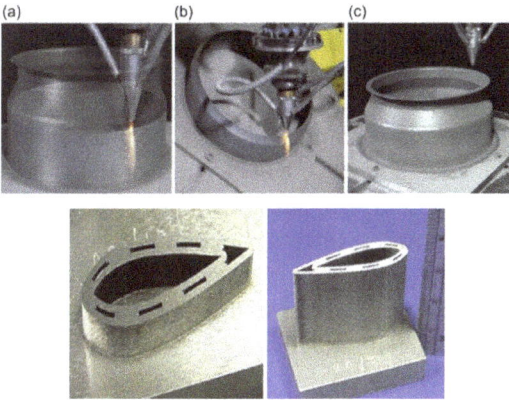

**Figure 28:** Laser-engineered net shaping (LENS®) of IN 718 helicopter combustion chamber (top) and Ti6Al4V airplane turbine blade (bottom) [915, 922].

## 3.3.2 Electron beam melting

Electron beam melting (EBM) is an AM process in which source of energy for melting is high-energy, localized and focused EB rather than laser. The main difference with EBM is that it can be applied only to metallic components since electrical conductivity is required. However, the advantages include that it can be moved at extremely high velocities (scan speed) and beam power is merely function of available power from grid.

Material properties initially were adversely affected by the amount of porosity generated during the process which now has been overcome to a large extent. Nowadays, treatments by highly focused and concentrated EB results in very finely dense structure whose properties are comparable to properties achievable by other processes such as casting and welding. In addition, there is high potential of the process – inherent, rapid and directed solidification which leads to very fine microstructure and epitaxial growth (a unique feature only possible in EBM) [874]. The history of AM goes back to the time when EB was harnessed to be used for imaging (microscopy) and subsequently some studies were reported in the United States in the mid-1980s [923]. They were not carried out with respect to using electron energy for part buildup. The objective was only to carry out melting using EB. Real progress in this field came in 1997 when Arcam AB Corporation in Sweden commercialized first selective EBM (SEBM) machine [873, 924, 925]. The process is fundamentally very similar to scanning electron microscopy (SEM) which involves generation of an EB by heated W filament, their collimation and acceleration to 60 keV (usually much higher than that in microscopy), and finally impingement on build table housing powder after passing through two magnetic focusing coil (magnetic lens) system. Beam current is controlled in the range 1–50 mA and the beam diameter ($\emptyset$) ~ 0.1 mm. Powder size typically ranges from 10 to 100 µm and typical layer thickness it forms on the bed is 0.05–0.2 mm. EB scans the surface in two passes. *First pass* consists of traversing path at higher speed (~10 m/s). This step is repeated multiple times and is carried out to preheat the powder to sintered state. This is followed by second slower scan (~ 0.5 m/s) during which melting occurs. Once top layer is solidified, a roller lay down another layer of powder and process is repeated. Entire process takes place under vacuum (typical $10^{-1}$ Pa in vacuum chamber and $10^{-3}$ Pa in electron gun). In addition to vacuum, a low-pressure inert gas helium is added to chamber at a pressure of $10^{-1}$ Pa to avoid buildup of charge of powder. Once part buildup is complete, part is allowed to cool inside chamber which may be assisted by further purging of He [926]. Generally speaking, the path electrons traverse to build up a surface can be categorized into three roasters: heating, melting: hatching and melting: quasi-multibeam. These are briefly described in Figure 29.

The process is virtually suited to all types of metals and alloys resulting in various types of metallurgy [927], melt pool geometry, its configuration, temperature distribution in it [67] and type of materials handled and processed [873, 874]. The process has been recently studied by the help of computational modeling and simulation for thermal heat transfer [928] and microstructural evolution [97, 472]. There have been few studies aiming at EBM of glass forming alloys in the early [929] and late 1980s [930], but they were primarily aimed at carrying out melting using electron energy rather than part buildup. Recently, few reports have been published about production of BMG by use of EBM [931, 932], out of which notable effort has been made at Mittuniversitetet, Sweden, in collaboration with Exmet AB, but it is

| (a) Heating | (b) Melting: Hatching | (c) Melting: Quasi-multi-beam |
|---|---|---|

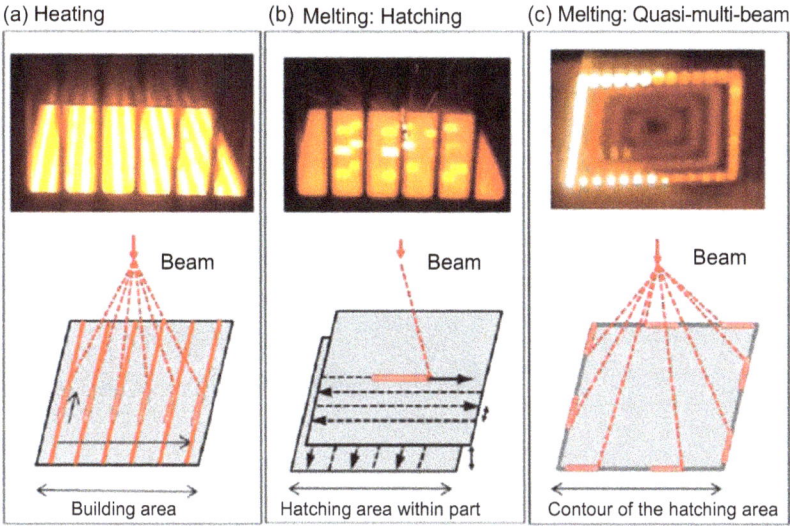

| Building area | Hatching area within part | Contour of the hatching area |
|---|---|---|

**Figure 29:** Actual pictorial (top) and schematic (bottom) representation of heating and melting during SEBM [874].

first of its kind and a lot of research and development effort is needed to success-fully produce large and complex BMGMC parts using EBM.

## 3.4 Characteristics of process

There are many unique characteristics of the AM process which give it a distinct po-sition and advantage over other manufacturing processes. Some of these include LBL formation, rapid cooling and in situ heat treatment of base layer. These will be briefly described here.

### 3.4.1 LBL formation

LBL formation is a unique feature of AM. It is the name given to evolution of unique solidified layers which are formed on top of each other as the process continues. It gives rise to layered structured deposition of metal/alloy in a built-up fashion. Usu-ally, it is associated with two parameters: (a) vertical build direction and (b) cross build direction (Figure 30).

Both have their own unique features/role in AM and effect on final alloy quality. Vertical build direction is associated with building of alloy in a direction normal to plane of rest or axis of table ($x$-axis). This involves deposition of metal layers on top of each other in such a way that finally a part is built which carries mass which is

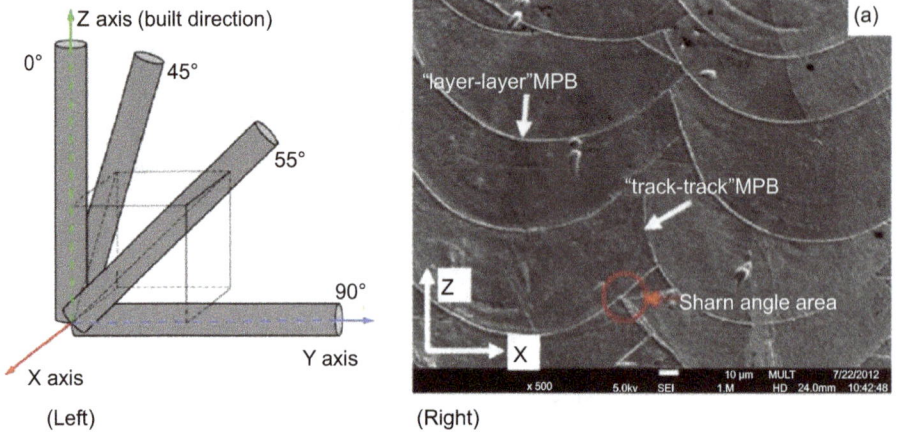

**Figure 30:** (Left) Schematic of build directions [933], (right) typical micrograph showing vertical build direction (Y), cross build direction (X) and their effects [934].

accumulation of welded/joined layers. As described in previous section, this built is highly a function of scan speed, laser energy, scan depth and material chemistry. The resultant part, its final chemical make-up and compositional homogeneity is dependent on how accurately and precisely aforementioned parameters are controlled. Excessive time spent at one point will result in localization of heat and may cause extreme deterioration which may include burning of part. This is very critical in low melting (Al and Mg based) alloys. In high melting metals (single component or two component (binary) alloys), this is not a big problem but in multicomponent alloys (BMGs, HEAs) this may cause loss in chemical composition due to burn out (oxidation) of low melting alloying elements. Other detrimental effects of this are: segregation, anisotropy, crystallographic distortion (elongation of crystal axis), improper and incomplete melting and fusion between layers, and formation of unwanted phases and intermetallics (IMCs) which can severely hamper the final alloy's mechanical properties. Usually, optimization in this direction is highly a function of operator skill and experience. However, a lot of help is provided in modern days by data and visual stimulated effects and graphics produced by modeling and simulation of various concurrently happening phenomena [893, 935] during AM. Buildup along $X$-axis (i.e., along the axis perpendicular to the normal built direction) is another important parameter which must be controlled properly, adjusted and optimized to get good quality part. Any improper functionality in this will result in improper fusion between adjacent layers which may give rise to porosity and undefined porous structure which not only weakens the final alloy but also is source of bad appearance and make it prone to atmospheric and weather (corrosion) attacks. On the other hand, use of very high laser power while building LBL pattern in adjacent layer will result in a fusion, melting and compositional variation at points of contact (tack points) between LBL of the first and second layers. This is very good example of tack-to-tack failure which is very common under fatigue loading.

### 3.4.2 Rapid cooling

Second most important feature of AM is rapid cooling. It results from very shallow depth and very narrow size (length and width) of melt pool. In addition to this, it has transient nature. That is, it keeps changing its shape and pattern of heat transfer from it during the time laser scans its path following CAD geometry. These features of melt pool make it highly unstable and it tends to quickly return to its stable form by dissipating heat away from it (solidifying). This gives rise to rapid cooling. This rapid cooling is particularly beneficial in multicomponent alloys where it facilitates texture development and directionality of properties. For example, in case of additively manufactured turbine blades, their properties are comparable (in fact superior) to directionally solidified blades by conventional means. This rapid cooling also reduces the fusion zone thus decreases the chances of spatter, heat-affected zone (HAZ), fuming and other detrimental welding effects. In the case of BMGMCs it is highly desirable as it helps in in situ glass formation which is highly desirable with recourse to an additional processing step and without adding to additional processing cost.

### 3.4.3 Assimilation of free volume

It is consumption of free volume during AM of BMG. It happens when material gets reheated to higher temperature after solidification from melt. Strained structure of glass undergoes various transitions such as relaxation and rejuvenation to form a final structure. This is accompanied with a large change in volume (decrease) when free space in between networked glass gets assimilated and finally a dense structure is formed which is free of volume. This process, however, results in overall shrinkage or decrease in volumetric space in material [936].

### 3.4.4 In situ heat treatment

Another very big feature of AM is in situ heat treatment of layer beneath fusion layer. As the primary layer which is raised to melting temperature ($T_m$) by tuning of laser power to high value, traverses its path following CAD geometry, it heats the layer beneath it which is already solidified. This heating is sufficiently high enough to cross first transformation temperature (lower critical temperature/invariant temperature) in most alloys. This triggers the first solid state transformation (solid–solid transformation) which usually is precipitation out of stable crystal forms out of meta-stable highly (nonequilibrium) cooled products in the first layer. This in situ heat treatment is integral part of most AM processes (at least those involving fusion (SLM/LENS®)). In addition to formation of stable crystal forms, it helps in homogenization, normalizing, improvement in mechanical properties and in some cases grain refinement of final alloy structure. In BMGMCs, this causes "devitrification" which is highly beneficial to increase toughness and ductility of this class of alloys. A noticeable disadvantage of this heat treatment is formation of IMC which may form as a by-product of solid–solid transformation reactions. In a worst-case scenario, incipient fusion/melting will occur at point where low melting point constituents exist resulting from introduction/inclusion of impurities from first high-temperature layer formation process. This not only can mechanically weaken the alloy but can also cause formation of porous network which is an additional impulse to decrease its quality. An effective way to avoid this heat treatment is to control machine parameters (laser power, scan rate, spot size, etc.) in such a way that only a necessary amount of heat is transferred from laser to the metal without causing development of zones of heating.

## 3.5 Bulk metallic glasses by additive manufacturing

BMG are distinctive from their composite counterpart in that 100% glassy or fully vitrified structure is desirable and achieved during the process. The process requires very good control on chemical composition and process parameters in that utmost dexterity is required. Any good glass former will result in formation of fully glassy structure but an excellent control in process parameters (laser power, size and scan speed) is required to avoid crystallization in layer beneath fusion as layer-by-layer formation is build. Part achieves consolidation only if excellent control is exercised and it is inherent feature of system and intrinsic property of material being processed. Optimization is critical and often process is more suitable for forming BMG matrix composites as explained in the next section.

## 3.6 Bulk matrix glass matrix composites by additive manufacturing

Processing of BMGMC by AM [59, 60] is slowly, progressively but surely growing as successful technique for their production on large scale. Various forms of AM processes (SLS, SLM/LENS® [777], DLD [482, 483], EBM) are slowly but surely attracting the attention of scientists around the globe to exploit their potential to be used as large scale industrial technique(s) for the production of BMG. Despite the inherent bottlenecks in the AM processes, there have been successful reports about their production preferentially by SLM – a form of AM involving complete fusion. Various types of glassy structures, for example Al- [467, 469], Zr- [69, 70, 84, 87, 88, 412, 475, 477–479], Fe- [62, 307], Ti- [844] and Cu-based [477] BMGMCs, have been successfully produced using SLM (AM).

It is well known that incipient metal fusion, its transience, progression (movement) and subsequent deposition out of melt pool following metallurgical principles (solute partitioning, alloy diffusion and capillary action to form dendrites) follow a LBL pattern. This LBL is distinct in almost all AM processes (SLS/SLM, DLD, EBM, etc.). The unique feature of this LBL pattern is that the top layer gets heated to metal/alloy temperature courtesy of tuning of laser power. This is called "fusion layer." Now, as the fusion layer traverses its path dictated by CAD geometry fed at back end (.stl file), it generates HAZ preceding its tip. This HAZ is very much similar to HAZ observed in other fusion welding processes. The metal following it is usually found in solidified fine equiaxed grains form. This tendency is good normal behavior of fusion layer and results in good glassy structure (high glass-forming ability (GFA)) in BMGs provided melt pool temperature is high enough to cause complete melting and heat is rapidly quenched out of it making a monolithic glassy structure. This results in hard brittle layer. Now, as the complete path in this layer is traversed, it is descended by few microns (dictated by initial alloy properties and machine parameters), and is supplied with new layer of metal/alloy powder by the help of scraper/roller. The laser again starts traversing the path fed to it in the form of sliced CAD pattern. This layer again reaches melting temperature and incipient fusion/melting takes place at laser/metal contact point. However, this time, a unique new phenomenon takes place. As the layer currently in contact with laser melts, it generates enough heat for the layer beneath it to reach a certain high temperature as well (usually 0.5 $T_m$ and $>T_x$). This heating of lower layer is enough to take the alloy back into nose region of time–temperature transformation (TTT) diagram which causes its crystallization (solid–solid transformation (devitrification)). Depending on the alloy chemistry and amount of time spent at temperature above $T_x$ (in nose region of curve), there could be (i) complete glassy structure, (ii) partial glassy structure or (iii) complete crystalline structure (no glass). Last is usually meant to be avoided during BMGMC processing and second is desirable.

There is, however, very narrow window of composition and temperature during which complete glass formation or complete crystalline structure formation could be avoided. (a) Only alloys with very high GFA should be selected from composition perspective and (b) should be tailored to cool with sufficient enough cooling rate (calculable from exact TTT diagram) which should cause their in situ equiaxed ductile phase dendrite formation during primary solidification in first layer retarding complete glassy state or crystallinity. Once in situ structure is formed, re-heating of the lower layer to a temperature in the nose region of TTT diagram during devitrification does not have much effect on further crystallization (due to kinetics (solute partitioning)) provided it should not be purposefully allowed to stay there for long. In general process, from a fundamental theoretical stand point, a 100% monolithic glassy structure or glassy matrix with fully grown in situ crystalline dendrites does not further undergo transformation to another crystalline phase (as they have already come out of their metastable glassy state). A powerful impulse on this could be caused by the introduction of carefully selected potent inoculants which are added to alloy melt during melting stage. These may serve as active nuclei for the preferential heterogeneous nucleation of ductile phase dendrites during primary solidification ensuring the least formation of metastable glassy state which in turn reduces the possibility of conversion of glass to crystallites during subsequent heating of layer (devitrification stage) as there is no glass (all the metastable or unstable phase have already been transformed to their thermodynamically stable state). No such effort has been made in the past to exploit this unique crystallographic feature of alloying in AM. This forms the basis of present research.

Few leading groups in the world have recently produced BMGMCs by AM. A brief tale of some of these is narrated here. Flores et al. [478, 479] successfully studied the effect of heat input on microstructure of Zr-based BMGs manufactured via LENS®. They observed the formation of unique spherulites within HAZ at high laser input ($10^4$ K/s) which disappeared as laser power is reduced (Figure 31).

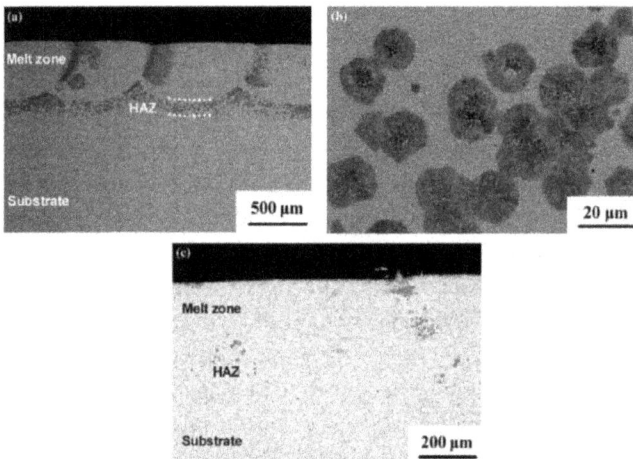

**Figure 31:** Cross-sectional backscattered SEM images of laser-deposited layers on the amorphous substrates processed at a laser power of 150 W. (a) and (b) Microstructures obtained at a laser travel speed of 14.8 mm/s. The featureless melt zone is shown in (a) surrounded by a crystalline HAZ, and the isolated spherulites of the HAZ are shown in (b). (c) Increasing the laser travel speed to 21.2 mm/s reduced the formation of the HAZ to only a few isolated spherulites [478].

These spherulites bearing unique crystal morphology seem to bypass isothermal cooling microstructures – a phenomenon not observed previously. The same effect was observed in their earlier studies on Cu-based BMGs [477]. In another study, a supervisor from the author's group (MAG) with coworkers [485] studied the effect of compositionally gradient alloy systems to manufacture BMGs and HEAs composite layers via LENS®. They aimed at finding an optimized composition at which effect of both alloy systems can be obtained in conjunction. Alloy systems consisting of $Zr_{57}Ti_5Al_{10}Cu_{20}Ni_8$ (BMG) to $CoCrFeNiCu_{0.5}$ (HEA) (first gradient) and TiZrCuNb (BMG) to $(TiZrCuNb)_{65}Ni_{35}$ (HEA) (second gradient) were used and processed at 400 W, 166 mm/s and 325 W, 21 and 83 mm/s, respectively. Using selected area electron diffraction patterns, they successfully reported the formation of fully amorphous region in the first gradient and amorphous matrix/crystalline dendrite composite structure (Figure 32) in the second gradient in individual melt pools.

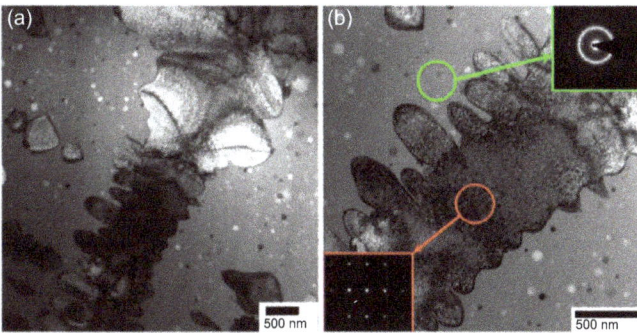

**Figure 32:** (a) TEM BF image of the laser surface melted region processed with a laser power of 325 W and a travel speed of 83 mm/s and (b) TEM BF image with the corresponding electron diffraction pattern of the crystalline dendrite (lower left inset) and amorphous matrix (upper right inset) [485].

Increasing the speed caused a slight variation in morphology and composition. Their results were consistent with their earlier investigations [324, 480]. However, the effect of reduced power and/or increased speed is needed to validate GFA of these systems. Zhang et al. [70] investigated the effect of laser melting in the form of surface remelting and solid forming on a well-known $Zr_{55}Cu_{30}Al_{10}Ni_5$ hypoeutectic system. They observed that despite repeated melting of alloy four times on its surface (LSM) during single trace, there was no effect on its glassy state. However, during solid forming (LSF), distinct crystallization was observed in the HAZ between adjacent traces and subsequent layers after first two layers. A series of phase evolutions were observed in as – deposited microstructure as it moves from molten pool to HAZ. In these microstructures, $NiZr_2$-type nanocrystals and equiaxed dendrites form from rapid solidification (L-S transformation) during LSM, while $Cu_{10}Zr_7$-type dendrites form as a result of crystallization of pre-existed nuclei (S-S transformation) in already deposited amorphous substrate. This paved the way for better understanding and application of LSM and LSF in terms of GFA and crystallization. Another group at the University of Western Australia led by Prof. T. B. Sercombe developed aluminum-based BMG by SLM [467–469]. They showed that an empirical laser power exists (120 W) at which width and smoothness of scan track are optimal; that is, defects (cracks (parallel, perpendicular and at 45° to scan track) and pores) in the scan edge are almost eliminated at this laser power. Crystallization, preferred orientation and melt pool depth are observed to have direct relationship with laser power while the pool width is observed in inverse relationship. Four distinct regions of scan track (fully crystalline (~100 nm), partially crystalline (~500 nm), boundary between amorphous BMG and bigger crystals and edge of HAZ (no crystal)) are identified. They further studied preferred orientation and found it to be a major effect of devitrification (both by very high laser power (pressure wave) and by temperature (oxidation)) as measured by EDS.

A few more notable studies have been reported very recently by leading research groups around the globe in which $Fe_{68.3}C_{6.9}Si_{2.5}B_{6.7}P_{8.7}Cr_{2.3}Mo_{2.5}Al_{2.1}$ (at%) [307], Fe-Cr-Mo-W-C-Mn-Si-B [937], other Fe-based BMGs [62, 891, 938], Ti–24Nb–4Zr–8Sn [473], other Ti-based BMGs [474], $Al_{85}Nd_8Ni_5Co_2$ [306], Al-based BMGs [939–942], Zr-based BMGs [69, 70, 84, 943, 944] and biomaterials and implants [833, 945] have been processed by SLS/SLM. Interested reader is referred to cited literature.

# Section 4
# Modeling and simulation

## 4.1 Why modeling and simulation?

Although in use since ancient Roman times [946], modeling and simulation picked up interest and achieved pinnacle in modern-day scientific and engineering sectors with the advent of computer technology which came not more than two decades ago. Now, it has proved itself to be an important and integral part of product and part design, product development as well as prediction, utilization and enhancement of properties. Various branches of modeling and simulation, ranging from part-scale modeling which involves development of codes of theorems in advanced computing platform such as Java®, C, C++ and MatLab Simulink® to their simulations in customized simulation packages such as Solidworks®, Ansys® and Catia® to performing complex atomistic simulations in dedicated software, have now become integral part of design procedure in major industrial clusters. Its use in research and development is also becoming an important part of the whole process to eliminate the so-called hit-and-trial" methods, which are not only waste of time but energy, materials and resources. In materials science and engineering mainly two of its branches are routinely used. These are "part-scale modeling and simulation" and "atomistic modeling and simulation." The former is used for the complete design of complex machinery segments, equipment, assemblies, subassemblies, their materials of fabrication and property prediction in different regions as a function of extrinsic parameters such as heat, velocity, pressure and time, while the latter is used for prediction, estimation and improvement in atomic-scale properties using theories of atomic configuration and arrangement mainly relying on intrinsic parameters (specific heat/latent heat, heat of fusion, etc.). The unique ability of atomistic modeling and simulation is that it uses atomic functions and their variables to generate knowledge about their behavior under various impulses. In both cases, the use of these methods is of big help and support in saving time, materials resources as well as improving the functionality and property enhancement.

## 4.2 Capabilities/powers and limitations

The exponential rise in the use of modeling and simulation with the advent and progress of computer technology and increase of computing power of machines gave rise to much easiness in the design and development process. Many difficult, or, in some cases, impossible to envisage problems can now be simulated using these computing platforms. These include simulation of water flow and its patterns in rivers and channels, simulation of interior of the Sun, stars and other heavenly

https://doi.org/10.1515/9783110747232-004

bodies, cosmic events, nuclear engineering problems and so on. However, despite these advantages, there are still situations and applications which limit the use of modeling and simulation techniques. These include unavailability of strong efficient computing algorithms (with lesser approximations) needed for the replication of actual real-world situations, unavailability of real-world experimental data (physical constants) needed to simulate a particular problems, unavailability of more accurate deterministic or nonprobability-based models using actual situations rather than basing their outcome on probability and so on. Owing to these reasons, there is still a need for further investigation and removing bottlenecks from modeling and simulation problems.

## 4.3 Types of modeling and simulation

Modeling and simulation techniques can be divided into following types depending on how the process is carried out and result attainment is sought out. The brief description is given below.

### 4.3.1 Energy minimization

In essence, energy minimization (EM), also known as energy optimization or geometry optimization, can be described as a set of numerical methods to find lower potential energy surface/state starting from state/surface of higher energy. These are extensively used in chemistry, physics, mathematics, metallurgical engineering, chemical engineering, mechanical engineering and aerospace engineering to find stable/equilibrium states of molecules, solid and items. Extensive studies have been carried out in various fields making use of EM techniques to formulate models highlighting the importance, significance and use of this method in modeling and simulation and solution of engineering problems. Its various submethods include the Newton–Raphson method, the steepest descent method, conjugate gradient methods and the simplex method.

### 4.3.2 Molecular dynamic simulations

Molecular dynamic (MD) simulation is another powerful method of solving physical movements of atoms and molecules. It heavily relies on Newton's equations of motion for solving a system of interacting particles in which their motion is calculated by interaction in a set period of time. Various authors have applied these techniques for solving various types of phenomena including heat transfer [947, 948], heat transfer coupled with phase change during laser–matter (LM) interaction [949], modeling and simulation of mechanical properties of polycrystalline materials [950], problems of hard and soft matter physics.

### 4.3.3 Monte Carlo simulations

Monte Carlo (MC) simulations are a set of methods that rely on random sampling to compute their results. They are often used in computer simulations of physical and mathematical systems. These methods are typically effective in calculating solutions of systems with many coupled degrees of freedom such as fluids, disordered materials, strongly coupled solids and cellular structures. Their method of operation is based on handling many systems at one time. However, they are not "many body systems" which are famous ways of solving atomistic problems. They differ from the latter in a sense that atomistic methods deal with solving potentials and functions of atoms and their electronic states which can be modeled using MC techniques [951, 952] but are not integral part of these calculations. Some of the prominent examples detailing MC methods are solutions of heat transfer problems [953–957], microstructure prediction [958], carbon materials [959] and nucleation phenomena [960]. Recently, they have also been applied to studies of bulk metallic glasses (BMGs) [961, 962].

### 4.3.4 Miscellaneous methods

There are further miscellaneous methods which are used for the modeling and simulation of various physical and engineering phenomena. Some of the prominent are Langevin dynamics (LD), normal mode (harmonic) analysis and simulated annealing (SA). LD methods are used for the mathematical modeling of dynamics of molecular systems. In summary, they can be explained on the basis of Newton's second law of motion by adding two force terms (one for frictional force and other for random force) to approximate the effect of neglected phenomena. Their examples are solution of protein folding problem [963], and BMGs [964]. *Harmonic analysis (HA)* are used for simulation in which the characteristic vibrations of an energy-minimized system and the corresponding frequencies are determined assuming its energy function is harmonic in all degrees of freedom. Normal mode analysis is less expensive than MD simulation, but requires much more memory. Similarly, SA is a random-search technique which exploits an analogy between the way in which a metal cools and freezes into a minimum energy crystalline structure (the annealing process) and the search for a minimum in a more general system; it forms the basis of an optimization technique for combinatorial and other problems [965]. Their use in BMG problems exists in a few cases but bears a prominent potential keeping in view the complexity of problems encountered in BMG systems.

## 4.4 Modeling and simulation of nucleation (microstructural evolution) in solidification

Modeling and simulation of nucleation and growth (NG, microstructural) evolution processes have been extensively studied since the early days of development of theories of modeling and use of computer technology for simulation. Initially, linear analytical/deterministic approaches were adopted to model simple transport phenomena pertaining to heat and mass transfer in one dimension only. Development of classical nucleation theory (CNT) [966] and its application to BMGs [967] is an excellent example about modeling of nucleation phenomena. They were quite useful in explaining the NG in one dimension. However, they suffer from three main drawbacks: (a) linear, one-dimensional models cannot explain actual nucleation phenomena happening in three-dimensional bulk of alloy liquid leading to evolution of solid in bulk; (b) these linear models are not good to explain the nonlinear nature of transient processes, especially heat transfer which is a strong function of changing process parameters itself (and the evolution of new front is dependent on new properties in the preceding step); and (c) the intrinsic nature of solidification process is probabilistic. Analytical modeling (even in its transient state) gives exact answers without taking into account the time-dependent probabilistic nature of process thus chances of error exist. All these reasons lead to the evolution of probabilistic modeling which was not only tailored according to the need and nature of process but also take into account the errors which might have been resulted from changes in system variables and were not solely based on permutations and/or combinations. These models were one dimensional in nature initially, which evolved into two and three dimensions progressively as their understanding increased. One of the first successful reporting of three-dimensional probabilistic model of nucleation dates back to 1993 at EPFL, Lausanne by Rappaz, M and Charbon, Ch [100] (Figure 33).

In detail, they presented two models simultaneously incorporating mix mode modeling (combining the powers of deterministic with probabilistic modeling) occurring at different stages of solidification throughout the cross section of alloy casting. One was aimed at explaining the phenomena in two dimensions [100] and other in three dimensions [101]. In this model, they used previously developed two dimensional models primarily aimed at solid state transformations and used their own approximations to model three-dimensional liquid-to-solid transformation phenomena (explained in detail in later sections). During the same time, another effort was made at Ecole des Mines de Nancy, France by Ablitzer, D [968] to model different transport processes occurring in different metallurgical unit operations. However, the limitation associated with this method was the use of linear deterministic approach, which was satisfactory only to a limited extent. Different probability-based methods (phase field (PF), cellular automation (CA) and front tracking method) evolved simultaneously (aimed at solving different problems, predicting grain size, determining nucleant size, dendritic arm size, interdendritic arm spacing) as

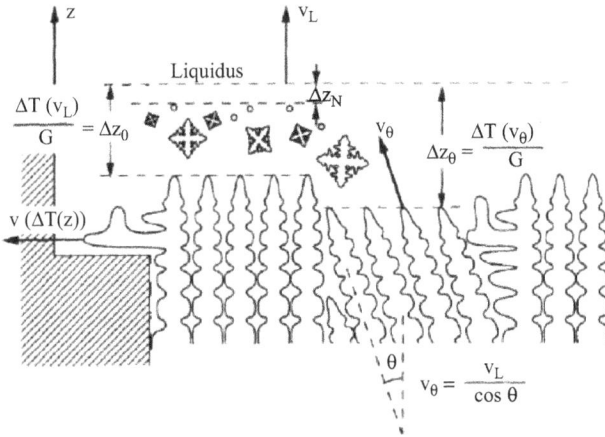

**Figure 33:** Schematic of various growth mechanisms occurring in a dendritic alloy with transient temperature conditions [100].

understanding of solidification process gets deeper and deeper but in almost all approaches, Monte Carlo–based simulation was applied to solve complex phenomena due to its rigor and power for solving multiphysics phenomena in liquids, solids and gases (as described in Section 4.3). Out of these, CA-based approach remained successful to a large extent. These picked up pace and become a topic of interest in the scientific community when it was coupled with finite element (FE) approaches giving birth to the so-called cellular automation finite element (CAFE) methods [105, 106, 969, 970]. Now, these are the most important topics of research due to their versatility and ability to model solidification processes to an extremely fine detail and at a scale which is very close to actual solidification patterns and microstructures observed in real alloys [971–975]. This will be explained in detail in later section (Section 5).

## 4.5 Large-/part-scale modeling

Large-/part-scale modeling involves modeling and simulation of phenomena happening at part level. It involves application of mathematical modeling to whole surface or volume of part. These usually include large and complex three-dimensional algorithms as applied to explain the evolution of certain phenomena (heat transfer, mass transfer, stress patterns and fluid flow) happening at the scale of whole body. These may include application of more than one algorithm at one time in conjunction with the other (multiphysics phenomena). It usually involves the division of whole part in certain specific small segments and then applying the model (carrying algorithm(s)) to each individual segment. All segments are then combined altogether to generate a synergic effect which gives rise to the evolution of solution of certain problem as a whole. This involves computational and/or analytical (numerical) integration. In essence, its description and detail are described below.

### 4.5.1 Analytical modeling

Analytical modeling usually consists of finding a numerical analytic solution of complex mathematical equations using standard mathematical or statistical techniques. These techniques further imply the use of conventional or (non-conventional) modern mathematical methods. Models generated using these techniques are usually one dimensional or extension of one dimensional to three dimensions using simple approximations. Further, these could be linear or non-linear. A computing platform is unable to read mathematical solution of, or even base, calculus equation. It is necessary to translate this equation to simple analytical solution using numerical methods. This gave rise to numerical analytical modeling. These include the Newton–Raphson method, the Runge–Kutta method and other approximation methods. There are further two types of analytical modeling: (1) deterministic/continuum models and (2) stochastic/probabilistic models.

**Note:** Use of "analytical" or "numerical" terms is dedicated to two different forms of solutions. The former is referred to finding a solution using standard mathematical equations such as those based on Newton's laws of motion, Einstein's theory of relativity, law of gravitation and so on. while the later is referred to finding the solution of complex calculus or partial differential equations (PDE) using more advanced or recent integration techniques which in essence translate an integer or PDE to an analytic form which a computer can understand such as the Newton–Raphson method and the Runge–Kutta method. However, sometimes, these are synonymously used as well. For the sake of this book, these are referred to as "numerical analytical model" for "analytical model" and "numerical probabilistic model" for "probabilistic models."

### 4.5.1.1 Deterministic/continuum models

Deterministic/continuum models are models in which no randomness is involved in the development of future states of the system. Thus, a deterministic model will always produce the same type of results irrespective of initial state or start point. In other words, a deterministic model will always produce exact solution. Most of the equations (PDEs) describing physical phenomena in mathematics and physics are deterministic in nature. A good example is Schrödinger equation whose solution is exact in nature despite the relation between its wave function ($\psi$) and properties may have probabilistic nature. A typical example of nondeterministic (analytical) model/equation is Heisenberg's uncertainty principle.

### 4.5.1.2 Probabilistic/stochastic models

Stochastic/probabilistic models are models in which little or some randomness is always involved which is depicted in the final result. Thus, a set of values and initial parameters will always lead to a group of outputs all, or some of which are true in those conditions. The examiner has the liberty of choosing any of them as per their requirement [976]. Probabilistic models are not exact but are based on best approximation. Based on the hierarchical design of computing memory and power (binary 0101-type sequencing), it is easy for the computer to understand these models. In other words, it can be said that probabilistic models are based on statistical analysis of a problem. Also, their dependence is largely based on the random change of states (i.e., in these models state function also keeps changing). Although this can be explained by deterministic models as well, it is much better to compute problems encountering a randomly changing state variable in terms of probabilistic models. They are best to explain complex phenomena as they divide the problem in to small group sets which eventually leads to group of results out of which best possible output could be generated. For example, application of wind force to tall structures is best described by probabilistic/stochastic models as the amount, type, nature and direction of wind load keep changing with time (time-based evolution studies). Similarly, earthquakes and design of earthquake-resistant structures are best explained by stochastic modeling.

### 4.5.2 Computational modeling

Computational modeling, as the name suggests, is a technique or a group of techniques in which rigor and strength of computing power are used to model and simulate a numerically translated analytic or probabilistic equation carrying an embedded model. Computational models and their use in predicting actual real-world behavior of complex systems are a much more effective way of solving or arriving at near-perfect solutions. Computational models combine the power of both deterministic and probabilistic models in a few cases (mix models) to yield a much better result. The only impulse behind using these models is that they effectively employ and can incorporate even tough complex nodes which can be solved only using superior data handling and rationalizing power of computers rather than the human brain. The latter might take months to solve an equation at all nodes of system while a strong computer require only a few minutes to arrive at the same conclusion. In addition to that, the ability of a computing cluster to handle complex multiphysics randomly occurring simultaneous (parallel) phenomena at one time gives rise to their popularity in modern-day science and engineering. Some good examples for these models are FE models (FEM), finite difference models (FDM), PF models and CA approach.

### 4.5.2.1 Lattice Boltzmann method (LBM)

The lattice Boltzmann method (LBM) involves solving basic "continuity" and "Navier–Stokes" equations and simulating complex fluid flow situations. LBM is based on microscopic models and mesoscopic kinetic equations. It originally has its roots from Ludwig Boltzmann's kinetic theory of gases. The basic idea behind these methods is that gas/fluid can be perceived as a huge volume consisting of randomly moving small particles. The exchange of momentum and energy between these is achieved through head-on, one-to-one collision (just like explained by Brownian motion). Thus, in some cases, these could be viewed as FDM for solving Boltzmann transport equations. In a best-case scenario, if proper operators are chosen, LBM can even be used to recover Navier–Stokes equations [977]. In metallurgy, alloy solidification or microstructural evolution problems it typically takes into account the fluid flow, distribution of liquid in micro channels, solute segregation based on its partitioning, dendrite evolution in three-dimensional volume, its morphological development, interdendritic arm spacing and distribution of liquid in that space. It typically considers a fluid volume element which is cohort of particles represented by a particle velocity distribution function for individual fluid component at each grid point. These particles are considered to be distributed under constraint conditions (i.e., under the action of applied force) and, as described earlier, are under free random movement (Brownian motion). Another condition governing their movement, existence and evolution is that time average motion of particles is consistent with Navier–Stokes equations. Based on these parameters, very recently some studies have been carried out to model alloy solidification behavior in various types of alloy systems under different conditions [978–983]. Majority of these studies yielded strong and powerful results as they are conducted in parallel sequence (employing parallel computing techniques) combined with three-dimensional coupling of other modeling methods such as CA [982]. For example, in a study conducted by Eshraghi, M. et. al [982] a new LB–CA model was created to simulate dendritic growth during solidification of binary eutectic alloys. The LB method was used to solve the transport equations and a CA algorithm was employed to capture the solid/liquid interface. The strength of the model was that it was able to capture evolution of thousands of dendrites in a macroscale domain with approximately 36 billion grid points in 1 mm$^3$ region. The parallel model showed a great scale-up performance on up to 40,000 computing cores and an excellent speed-up performance on up to 1,000 cores. This was truly a great achievement in terms of modeling complex phenomena in large-scale macroscale domains as it bears a potential to be adopted for and applied on to massively parallel supercomputers (e.g., Jaguar at ORNL, Deep blue at IBM).

### 4.5.2.2 Phase field methods

PF methods are another most important class of computational methods which are used extensively for modeling and simulation of alloy solidification, microstructural evolution and associated transport phenomena in various types of alloy systems under variant solidification and processing conditions. These have also emerged as important simulation tool in recent years with the multifold advent of computers and increase in their ability to run various types of multiscale programs at one time (parallel programing). Following are some of the salient features of PF methods;

1. These methods revolve about finding a solution of PF parameter $\phi$, which is a function of position and time; to describe weather material is liquid or solid.
2. The behavior of this variable is governed by an equation that is coupled to equations of heat and mass (solute) transport.
3. The best way to explain "solidification pattern development" and "microstructural evolution" is to describe the evolution of liquid–solid interface in terms of smooth but highly localized changes of this PF variable $\phi$ between fixed values of 0 and 1 that represent solid and liquid.
4. One of the most powerful insights to the use of PF models is that it eliminates the use of boundary conditions. In normal solutions of problems, these boundary equations are required as they explain the evolution of liquid–solid interface in a particular confined volume in terms of solutions of transport equations (i.e., use of heat and mass (solute) transport by obeying thermodynamic and kinetic constraints). In PF calculations, these boundary conditions are not required. In these methods, the location of interface could be obtained by solving for numerical solution of PF variable at a position where $\phi = \frac{1}{2}$.
5. Another big feature of PF methods is that they treat topology changes such as coalescence of two solids regions that come into close proximity [94].

In detail, PF methods can be classified into various intersecting classes: (a) those that involve a single scalar order parameter and (b) those that involve multiple order parameters. Another way is to divide them into those (a) derived from thermodynamic formulation and (b) those which are derived from geometrical arguments. Their elaborative discussion and explanation could be found in latest literature cited [94, 95, 98, 984, 985]. They have also been applied to solidification and microstructural evolution problems in aluminum base [986] and general [987] multicomponent alloys and glass-forming systems [99]. PF methods (PFM) are rival to CA methods [988]. Both can be used in conjunction with each other. However, their real strength lays in their use with complex FE methods forming PFFE (phase field–finite element) or CAFE which will be described in a later section.

### 4.5.2.3 Cellular automation method

CA is the most important, powerful and latest development in computational models applied to solve the problems of transport phenomena happening in physical, chemical, metallurgical and engineering systems. In some cases, this is called third-generation modeling approach. In essence, it involves dividing the bulk volume of system (material) into a small number of finite cells, at the corners and boundaries of which the property of state variable and functions remain same/constant. Some of the salient **features** of CA method are (a) it is based upon strict physical mechanism (involving physical properties), (b) it makes use of low computation cost (thus benefits form low running cost), and (c) it can be coupled with other models (explained below). A **limitation** could be it may incur high initial (capital) cost (i.e., it may involve use of computing machinery with large processing and floating (RAM) memory).

The first successful CA model was presented by Rappaz, M and Charbon, Ch at EPFL, Lausanne in 1993 [101]. This model was built upon previous two-dimensional approaches of modeling solidification processes (primarily heterogeneous nucleation) and microstructural development and was extended in three dimensions by incorporating additional physical parameters. The power of model also came from the fact that it takes into account the crystallographic orientations as well. Essentially it comprises the following steps [100]:

1.  Grains and their properties (especially location and crystallographic orientation) are chosen randomly among a large number of cells and a certain number of orientation classes, respectively. For example, in the case of pure cubic metals (e.g., B2 CuZr equiaxed dendritic phase in glassy matrix in Zr-based BMG and their composites (BMGMCs)), the parameters of interest taken into account were; growth kinetics of dendrite tip and preferred (100) growth direction (which is already a well-known direction of heat flow and thus growth in these crystals).
2.  Further modeling then involves extending the application of initial model (which was applied to a small-scale region of solid) to whole volume of casting (solidifying melt). In other words, the transport equations that were applied first to a small finite solidifying volume are extended to the whole bulk by applying them step by step to a large number of small regions covering bulk volume.
3.  Constant temperature (usually ambient ($T_r$)) may be selected for a steady-state case.
4.  Initial temperature ($I_o$) may be selected for transient case.
5.  Grain growth morphology affects
    a.  columnar-to-equiaxed transition,
    b.  columnar zone morphology (which consists of selection and extension of columnar grains forming columnar zone) and
    c.  impingement of equiaxed grains (giving rise to their size reduction and fineness) is also taken care of.
6.  Final check with respect to effect of the alloy concentration and cooling rate which affect the final evolved microstructure are verified by rigorous experimentation.

The model has been specially used by combining it with FEM (CAFE) [104–106, 969, 970] as well as individually for the explanation of various metal and alloy systems (e.g., Mg alloy AZ91 [971], brass [107], Al–Cu systems [988], ordinary binary alloys

[972], Fe-based multicomponent alloys [989], TC4 alloy [973], hexagonal crystals [975] and other multicomponent systems [974]). The method has also been used for prediction of alloy microstructure in a specific process such as directional solidification [108], laser deposition [973], continuous casting [371] and additive manufacturing (AM) [112]. A powerful advantage could be achieved by combining CA with PF methods, but this approach is still in its nascent stage [975].

### 4.5.2.4 Miscellaneous methods
These include methods which employ techniques which involve use of special algorithms and computing strategy. They have their own advantages but with respect to complexity, mass volume of castings handled, and type, shape, size and morphology of melt pool encountered in laser assisted and other AM (electron beam) processes, these are rarely employed. Their special features and tailored abilities can be used in special circumstances and solving special problems. Some of the examples of these are virtual front tracking (VFT) model and sharp interface (SIF) model which are explained as follows.

### 4.5.2.4.1 Virtual front tracking model
This is quantitative model which has a specialty that it can be used to explain dendritic growth in systems which have low Péclet number (ratio of thermal heat convected to fluid (externally) to thermal heat conducted within fluid) systems. It is best when it describes thermal transport in two dimensions. In this case, it can even handle complex multiphysics fluid with high viscosities (liquid melt with pronounced mushy zone) [990]. Its variant (three-dimensional model) exists but computation becomes more complex and lengthy putting load on computing machinery and its power. A solution to this problem may be achieved by employing them only to less viscous (gas and liquid (petroleum-based)) systems [991, 992], but this is beyond the scope of solving for metallurgical problems (not the aim of this research). An excellent example of the use of this method is given by Zhu and Stefanescu [990], in which they solved solutal transport equation to generate data obtained by local temperature, curvature and local actual liquid composition to arrive at the difference between the local equilibrium composition necessary to describe kinetics of dendritic growth. By employing this strategy, the dynamics of dendritic growth (including noise-free side branching) from initial unstable (fluctuating) stage to steady-state stage was accurately predicted. Biggest advantage of adopting their technique was that it made the model completely mesh less when going for simulation in a computing platform. Further efficiency in decreasing the computational time was achieved by calculating dendritic growth directly from fraction solid (rather than initial growth velocity calculations). This was a big achievement and a huge advantage of using directly the thermal transport (mass transfer) data to arrive at geometrical features of growing interface (dendrite). Another excellent

example of coupling is given by Song, K. J and co-workers [993] in which they combined CA model with VFT to describe crystallographic preferred orientation-based β to α phase transformation of TA15 alloy at constant temperature (isothermal). They used both techniques to achieve solid-state transformations in this multiphase alloy. This again is excellent way forward to use various powers of different models to arrive at solutions of complex phenomena. However, more research is needed to optimize this regime.

### 4.5.2.4.2 Sharp interface model

These models again, as the name suggests, is a method or a set of methods which incorporate interface phenomena and evolution of a certain ("liquid–solid" or "solid–solid [994]") interface as a function of time. Owing to these reasons, these can best be used to explain evolution of interfaces during solidification as well as during heat treatment/solid–solid transformations. However, they have unique features of incorporating various auxiliary effects (surface energy, sharp moving interface, etc.) while aiming at the solution of final equilibrium state and its growth. Vermolen and coworkers [995, 996] used this method to solve diffusion equations for dissolution of second phase solid particles in multicomponent alloys taking account of surface energy effects. This was a unique work as this was the first attempt of its kind to explain dissolution in multicomponent alloys and was successfully explained by the use of SIF Model. They also described various other modeling suites developed previously which base their simulation strategy on SIF models and their improvement, namely, models by Argon et al. [997], Caginalp et al. [998], DICTRA (which simplified the complex problem of boundary conditions at the interface to a hyperbolic relationship) and so on. Recently, these models have also been applied to solve nucleation and ramified growth problems (NG) in electrodeposition [999] by using a combination of continuum and random noise (perpetuating SIF) terms in final equation. This again is an excellent example to combine the powers of two different modeling regimes. This trend is proving successful and is on the rise to solve difficult problems in complex situations [1000, 1001].

### 4.5.2.5 Mix mode modeling (analytical + computational models)

These topics have already been introduced in previous sections [105, 106, 970, 993, 1000, 1001], so it will not be described in detail here. Merely, an introduction will be given to create a feel of their operation and applicability. In essence, mix mode modeling and simulation approach is based on combining the powers of two different models under one umbrella. This necessarily does not involve their computation in parallel with each other in a so-called *parallel programming* on computing clusters (also known as supercomputers). However, if the latter is done, it yields much better solutions of complex problems in much less computing time and using much less computing power than what would have been possible if the solution is run in standalone mode on single computing machine. Essentially these can be divided into two main types: (a) *deterministic PF models* and (b) *deterministic CA models*. Their detailed description depends on individual branch of engineering or physical science to which they are applied and on the researcher (or research group) that is/ are responsible for their development. Popular regimes which now have become accustomed to be used toward solution of most of engineering problems related to multiscale/multiphysics transport phenomena are CAFE modeling, PFFE modeling and CA–VFT regimes.

## 4.6 Atomic-scale (atomistic) modeling

Unlike, part-scale modeling, another most powerful yet versatile method of predicting properties, phase change, their evolution and behavior is by the use of *atomic scale* or popularly known as *atomistic modeling*. Atomistic methods involve use of special theorems developed for solving atomic-scale functions of all atoms in a given volume of material. Their functions are solved in terms of individual as well as interacting potential of each atom. This gives final results representing properties of materials in terms of solution of their functions. Atomistic modeling techniques use modern superior computing powers to explicitly arrive at the combined solution of every atom in system. This in detail can be done by using various techniques specific to atoms or set of atoms (e.g., MD, MC, first principle (ab initio), force fields (interatomic potentials) with or without approximations). As interacting atoms are the foundation stone of all the materials, atomistic modeling has helped enable a new field of determining the properties of materials known as "computational materials studies." Some of the **advantages** of atomistic modeling and simulation are (a) they greatly help in reducing the cost of experimentation, (b) a huge saving in use of materials may be achieved, (c) process efficiency may be increased (d) overall time in running experiments (down time) is reduced. These abilities and advantages help in (a) discovery and (b) development of deeper understanding of varied, difficult and unexplained phenomena of materials. Despite this, one of the biggest **challenges** posed to atomistic modeling and simulation is an inherent difficulty to handle multiscale, multiphysics phenomena which span over many atoms (of changing wave functions) at one time. Another important area in which atomistic modeling has potential use is modeling with effective interactions between atoms, called interatomic potentials. These potentials do not treat the quantum nature of electrons explicitly; thus, models employing them are extensively faster than quantum-based models. They might suffer from loss of some accuracy but still can model billions of atoms at once. Good examples of use of such techniques are nanocrystalline materials, and friction between surfaces.

### 4.6.1 Classical molecular dynamics

As the name suggests, MD calculations are done to understand the dynamics/motion of assemblies of molecules in terms of their structure and the microscopic interactions among them to understand their properties at very small scale. A guess is developed at atomic level between interacting molecules and is presented to accumulate an effort to obtain exact predictions of bulk properties. History of MD techniques and their development dates back to 1957 when Alder and Wainwrigth first introduced the concept of interacting hard spheres and their phase transitions [696]. This was based on the original work to develop MC methods in 1953 (next section). A unique feature of MD techniques is that solutions generated are exact in nature such that they can be made as accurate as desirable. Only limitation is posed by computing power which is associated with capital and running budgets. Another feature of these calculations is hidden dynamic details behind bulk measurements can be revealed simultaneously without lapse in time step. An example of MD is the link between the diffusion coefficient and the velocity autocorrelation function, with the latter being very hard to measure experimentally, but the former being very easy to measure. Ultimately, the aim is to arrive at direct comparison with experimental measurements made on specific materials, making it absolutely essential to have the explanation of phenomena in terms of good model of molecular interactions. The overall aim of MD is to reduce the guess work and fitting to arrive at ensemble of minimum. In addition, MD techniques can also be used to explain difference between good and bad or just to generate a comparison based on generic data in which case it is not absolutely necessary to have a perfectly realistic molecular model but simple assumptions based on fundamental physics suffice. Very recently, MD simulations have been used for modeling of atomic-scale phenomena and BMGs spanning from understanding of structure [158, 1002], explanation of indentation-based deformation in bulk [341, 460], transition from elasticity to plasticity [1003], brittle fracture [1004] general mechanical (stress–strain behavior) properties [950], to evolution of structure in LM interactions [949].

### 4.6.2 Monte Carlo simulations

MC techniques are another set of procedures used for explaining atomic-scale phenomena in terms of set of probability-based possible outcomes out of which one could be chosen at random. History of MC techniques evolved with MD techniques and can be dated back to 1953 when Metropolis, Nicholas and coworkers at Los Alamos National Laboratories developed a method for solving state equations (encompassing interacting molecules) by fast computing machines [1005]. It was later modified in 1957 by Wood and Jacobson again at LANL [1006] and its implications are still going on in modern-day science and engineering. The power of MC calculations is that it can explain various types of phenomena and processes. For example, it can be used to describe and account for *sampling* from a given bulk. The objective in this kind of approach is to gather information about a random object by observing many realizations of it. In this kind of scenario, random physical mimicking is done to observe the behavior of some real-life systems such as production line or network of conglomerated materials. Another example of the use of MC simulations is *estimation*. In this kind of situation, MC techniques can be used to account for certain numerical quantities related to simulation model. An example in the artificial context is the evaluation of multidimensional integral via MC techniques by writing integral as the expectation of random variables. Yet, another most powerful insight to MC techniques is their use for *optimization*. This perhaps is the most powerful, most varied and most frequently used feature of MC techniques. In many applications, these functions are deterministic and randomness (which is a feature of probabilistic techniques) is introduced artificially in order to more efficiently search the domain of objective function. In a best-case scenario, MC techniques are used to optimize *noisy* functions, where function itself is random – for example, the result of MC simulation. This is the most varied feature used in solving materials science and engineering problems apart from *Sampling* [1007]. For the first reported time, It was used to explain phenomena and structure development in binary Ni–Nb BMG in 1993 [1008]. Very recently, it has been extensively applied to the problem of bulk metallic glass and their structural and mechanical evolution. For example, Hu et al. [1009] used this technique to describe probabilistic dual-phase magnetic behavior of Nd-based BMGs. Similarly, it is used to describe nanoscale structure development and structural relaxations (devitrification) in Zr-based BMGs [574]. A similar phenomenological study was carried out in Ni–Ti–Mo-based BMGs [1010]. In another study, MC techniques are used to study surface modification of Zr-based BMG by use of low-energy Ar or Ca ion implantation for biomedical application [1011].

### 4.6.3 Ab initio methods/first principle calculations

Ab initio methods or first principle calculations are set of methods derived out of quantum chemistry which are used to determine the state of a given system using its functions. Its features are based on quantum molecular orbital (MO) theory. It does not rely on empirical values of system but it is based solely on established laws of nature [1012]. Solution of Schrödinger equation forms the basis of MO theory. It yields energies and orbitals of electrons. *Energies* can be used to determine or expressed in terms of geometry optimization, reaction energetics, activation energies for kinetics and UV/Vis absorption prediction, whereas *orbitals* can be used for graphical display (including assessment where reactant might attack), charges, dipole moments, electronic potentials and NMR shieldings. In some cases, this is synonymous with MO calculations. History of these methods goes back to 1950 when Robert Parr first introduced their use to solve MO calculations of the lower excited electronic levels of Benzene [1013]. Since then, scientific world has seen many folds increase in their development. Most of these developments came as a result of (a) dramatic increase in computing speed as well as due to (b) design of efficient quantum chemical algorithms [1012].

### 4.6.3.1 Advantages
- They are easy to perform.
- Calculation cost is less (calculations are cheap in terms of their operation). However, experimental cost (to prove them) is high.
- Calculations can be performed on any system (even those that do not exist) while experiments are limited on only highly stable molecules (mostly monovalent or divalent).
- Calculations are safe (as merely, it is a mathematical procedure involving paper and pen or at most a computer) while setting up an experiment is a very risky and dangerous task (installation, operation, control and maintenance of large number of parts, equipment and machinery).

### 4.6.3.2 Disadvantages
- Calculations are too easy to perform by use of software (rendering them less use of human ingenious).
- Time to carry out calculations can be very lengthy (especially if used on low-power computing machinery).
- Calculations can even be performed on hypothetical (imaginary) systems (which are waste of time).

### 4.6.3.3 Procedure to carry out *ab initio* calculations
In essence, an ab initio calculation involves selection of (a) a method and (b) a basis set. Some of the examples of **methods** include
a. Hartree–Fock (HF) method and Slater determinant,
b. Car–Parrinello method and
c. Density functional theory (DFT).

For example, *HF method* is a simplest wave function–based method. It forms the foundation for more elaborative electronic structure method (SCF (self-constrained field), DFT, etc.). It posits a trial form of the N-electron wave function and uses the variational theorem (which is a wave function–based approach using mean field approximation [1012]) to obtain an approximate solution. The basic idea here is to express the wave function as a product of individual spin orbital solutions, the so-called *Hartree product*. This form of trial wave function does not obey the indistinguishability requirement for fermions that the swapping of two particles generates an identical but negated wave function: since in general all of $\chi$ functions can be different. In order to solve this enigma, instead of Hartree product, a similar multiplicative combination of individual spin orbital that forms the negative upon swapping is used. For an arbitrary number of electrons, this can be achieved using determinant of a matrix, called *Slater determinant*. Similarly, **basis set** (which is second part of performing ab initio calculations) is a set of atomic orbitals (AO), whose coefficients are to be determined using *HF* method to make MOs.

### 4.6.3.4 Approximations

Schrödinger equation can be solved exactly for H atom only to find solutions of MO theory (in terms of AO). So, all other AO solutions are approximations. Some of these approximations are

- *Born–Oppenheimer* approximation
- *Independent electron* approximations
- *Linear combination of AO* approximations

### 4.6.3.5 Evolution of HF theory

With the passage of time, HF method evolved into more simplified methods to find solutions of MO theory. In essence, they all involve solutions of time-independent Schrödinger wave equation which may be described as

$$\hat{H}\psi = E\psi \tag{6}$$

where $\psi$ is the wave function. It is a postulate of quantum mechanics and a function of the positions of all fundamental particles (electrons and nuclei) in the system. $\hat{H}$ is the Hamiltonian operator which is associated with the observable energy. It contains all terms that contribute to the energy of system

$$\hat{H} = \hat{T} + \hat{V} \tag{7}$$

where $\hat{T}$ is the kinetic energy operator and $\hat{V}$ is the potential energy operator [1012].

$E$ is the total energy of system. It is the operator associated with the observable energy.

These evolutions are briefly described as follows:

- SCF method evolved as a result of conjunction with the HF method. Thus, it is sometimes called HF–SCF method.
- Møller–Plesset (MP) perturbation (MP1) theory was the next evolution of HF theory. In this, Hamiltonian is divided into two parts: $\hat{H} = \hat{H}_0 + \lambda\hat{V}$. The second term is perturbation and is assumed to be small. The energy and wave function are expanded as power series in $\lambda$ (which is later set to unity). (Note: Even in solving MP1 theory, wave function and energy are HF wave function and energy.)

- MP2: It is the next evolution of MP1 theory in which wave functions remain the same. However, energy is explained in terms of MP2 energy and bears accumulative treatment of wave function by use of summations and so on.
- Since 1996, even MP2 method evolved into a more simplified theory known as DFT. The strength of DFT method is that the energy of system is obtained from electron density rather than from more complicated wave function. It is an approximation but is very good to explain the energies of system.

Recently, this method has been extensively applied to solve various types of problems in chemistry, physics, materials science and engineering. They have also been used to solve problems of BMGs and BMGMCs, in particular to study their structural evaluation (liquid state, its arrangement and final configuration into glass) [158] and origin of glass-forming ability both in binary and in ternary BMGs [1014].

### 4.6.3.5.1 Density functional theory

DFT is an evolved HF theory as explained in the previous section. It is required and thus developed to solve more complicated phenomena using simplified procedures and methods. It is a combination of simplified methods to find solutions of MO theory involving solutions of time-independent Schrödinger wave equation. As described in the previous section, it has found and is finding and establishing its footprint in solving BMG problems from atomistic point of view [699, 1015, 1016]. It is very helpful to solve problems such as state of system, alloying [1017], plasticity, NG, shear bands and their evolution, serrations (Section 4.11.2.2), deformation [1018], fracture and fatigue, structural, mechanical (hardness [1019]), thermal and electronic properties of metallic glasses [1020].

### 4.6.4 Interatomic potential/force fields

Despite their rigor and power, MD and MC methods cannot be used directly on atomic systems to explain their interactions. Their application requires the adoption of some rules that govern the interaction of atoms in a system. It is convenient in classical as well as semi-classical simulations to express these rules in terms of potentials. These potentials are known as interatomic functions. Usually, they are represented by a potential function $U$ or

$$U = U(r_1 + r_2 + r_3 + r_4 + \cdots + r_N) \tag{8}$$

where $U$ is the sum of potential energies of system of $N$ atoms. It is dependent on individual coordinates of atoms represented by $r_1, r_2, r_3, \ldots, r_N$. As explained in the previous section, some approximations are helpful in defining these potentials. For example, it is assumed that electrons adjust to new atomic potentials much faster than the motion of atomic nuclei (Born–Oppenheimer approximation). Similarly, forces in MD simulations are defined by potential $\vec{F}_i$ which is explained as follows:

$$\vec{F}_i = -\vec{\nabla}_{r_i} U(\vec{r}_1 + \vec{r}_2, \ldots, \vec{r}_N) \tag{9}$$

In general, selection of a particular type of potential depends on the following characteristics: *accuracy* (it is a measure of systems how well it can reproduce properties of interest closest to the required/desired values), *transferability* (it accounts for the ability of a system to allow the study of properties and generate results for which it is even deemed unfit) and lastly *computational speed* (it defines how fast calculations can be done on a system to solve its potentials). Finally, the choice of interatomic potential depends on the area of intended application. There are almost no "good" or "bad" potentials. Their choice and selection depends on the type of problem in question. On the basis of this, they can be "appropriate" or "inappropriate." High accuracy is often the prerequisite for problems pertaining to computational chemistry while computational speed is often a bottleneck in materials science, in which case processes have a collective character and require a large time for their effective solution.

### 4.6.4.1 Classification of interatomic potentials

Interatomic potentials can be classified into two types on the basis of how they will be used for a particular problem. These are:
- Pair (two-body) potentials and
- Multi (many-body) potentials

*Explanation:* Total energy of a system of N atoms with interaction described by an empirical potential can be explained in a many-body expansion:

$$U(\vec{r}_1, \vec{r}_2, \ldots, \vec{r}_N) = \sum_i U(r_i) + \sum_i \sum_{j>i} U_2(\vec{r}_i, \vec{r}_j) + \sum_i \sum_{j>i} \sum_{k>j} U_3(\vec{r}_i, \vec{r}_j, \vec{r}_k) + \cdots \tag{10}$$

Where

$\quad U_1$ – one-body term, due to an external field or boundary condition (wall of container).

$\quad U_2$ – two-body term, or pair potential. (This interaction of pair of atoms depends only on their spacing and is not affected by the presence of other atoms.)

$\quad U_3$ – three-body terms arise when the interaction of a pair of atoms is modified by the presence of the third atom.

This leads to classification of a system into two types: (a) pair potential (when only $U_2$ is present or calculation of potential is restricted to the second term only) (Figure 34) and (b) many-body potentials (when calculation of potential involves use of $U_3$ or higher terms). This classification also makes it easier for the procedure by which potentials are chosen. It is briefly described below:

1. *A functional form for a potential function could be assumed based on which certain parameters are chosen to reproduce a set of experimental data.* This gives rise to so-called **empirical** potential function. Most of the time, these are two body or pair potentials (e.g., Lennard Jones potentials, Morse potentials and Born–Mayer potentials).
2. *Electronic wave function for fixed atomic positions could be chosen.* This inherently is a difficult for a system of many atoms. Different approximations (as described earlier) are used, and analytical **semi-empirical** potentials are derived from quantum mechanical arguments. These typically are multi (many)-body potentials. (e.g., embedded atom method proposed by Baska, Daw at Sandia National Laboratories in 1984 [1021] and explained for fcc metals and alloys in collaboration with their colleague Dr. Foils in 1986 [1022] and cubic metals in 1992 [1023], glue method [1024], effective medium theory, reactive potentials, bond order potentials (TB potentials) [1025]).
3. *Directly performed electronic structure (quantum mechanics based) calculations of forces during so-called ab initio MD simulations (as explained earlier) could be chosen* (e.g., Car-Parrinello method using plane wave pseudopotentials).

Usually, *pair potentials* are performed for insert gases, intermolecular van der Waals interaction in organic materials and investigation of general classes of material

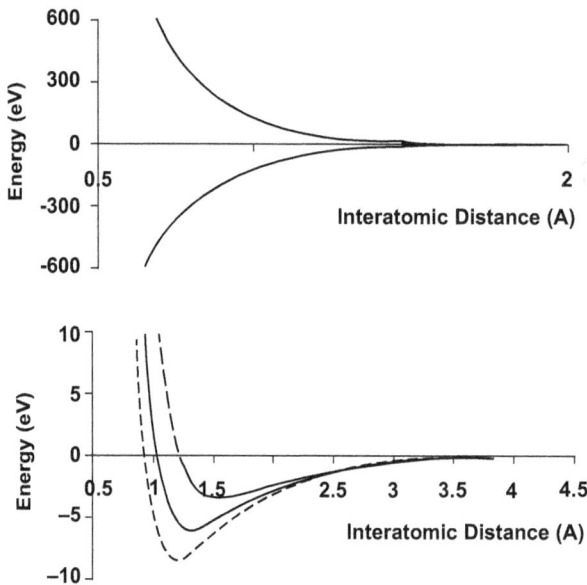

**Figure 34:** Plots of pair potentials: Top: attractive and repulsive pair terms as a function of interatomic distance. Bottom: pair terms for triple bonds (dotted curve), double bonds (solid curve), and single bonds (dashed curve) obtained by multiplying the attractive pair term by appropriate bond order value and adding it to repulsive pair term [1025].

nonspecific effects, whereas *many (multi)-body potentials* are performed for metallic systems. In addition, all other types of potentials have their own specific applications and are suited for the solutions of particular system (e.g., force field method can be used for covalently bounded systems, and reactive potentials are for carbon and hydrocarbons as described by Brenner in his original derivation [1025]).

## 4.7 Very recent trends/future outlook

An interesting (not yet adopted) approach would be to combine the power and rigor of both part-scale and atomic-scale modeling and simulation methods. This might involve the use of very powerful computing machines and strong parallel programing cluster, but it is bound to yield unique and fascinating properties of materials as they are used in actual application which otherwise are not possible as atomistic models are limited to be applied to confined volume of materials (usually hundred or few hundred of atoms) while part-scale simulations are not concerned with phenomena happening at atomic or molecular levels. They only yield data and predict macroscale behavior of part (and hence its material) as a function of external impulse (temperature, pressure, stress, etc.). In this regard, few efforts have been made to model the behavior of biological molecules (proteins) [1026], combination of transistor and its tunneling effect to form the world's smallest transistor at Oak Ridge National Laboratories' (ORNL) Jaguar Supercomputer [1027], multimillion atom biological MD simulation [1028] and so on, but still research in this field is very much in its nascent stage and restricted to large-scale well-developed computing clusters. More effort is needed to bring this technology to public and for its commercialization.

# Section 5
# Modeling and simulation of solidification phenomena during processing of BMGMC by additive manufacturing (AM)

## 5.1 Introduction

This section deals with evolution of microstructure during processing of bulk metallic glass and their composites (BMGMC) in incipient transient liquid melt pool formed in additive manufacturing (AM). Analysis is divided into two sections. The *first section* deals with evolution of melt pool as a result of interaction of highly localized, focused laser light with matter (metal powder). This results in the formation of melt pool whose shape, size, geometry and transient behavior are very much a function of heat transfer coefficients (HTC) evolving at every step of its formation (melting and homogenization) and dissipation (solidification). Solidification in this section is considered by modified general (classical) nucleation theory (CNT). Once formed, this pool travels as laser traverses its path all along the powder bed dictated by CAD geometry at the back end. The *second section* deals with microstructural evolution during solidification which is primarily *solute diffusion*– and *capillary action*–dominated phenomena. This is dealt with by microscopic two-dimensional (2D) and three-dimensional (3D) probabilistic cellular automaton (CA) models which model nucleation and equiaxed dendritic growth, resulting in the formation of microstructure in liquid melt pool as it solidifies (note: only "vitrification (glass formation)" effects are taken into account). The evolution of microstructure is checked against variation of number density, size and distribution of ductile phase in glassy matrix. Inoculants for ductile phase formation were selected previously by edge-to-edge matching (E2EM) [834, 835].

## 5.2 Modeling and simulation of heat transfer in liquid melt pool – solidification

As microstructure formed during selective laser melting (SLM) is mostly columnar [1029], it is a good indicator that heat flux transfer from melt is highly unidirectional; thus, heat transfer from bottom is transient one-dimensional (1D) process. Although heat is lost from material in the $x$–$y$ plane, that is, perpendicular to $z$-direction (perpendicular to build direction), its contribution is so low that it can be safely ignored. However, this was an old concept. New experimental observations have proposed new concept according to which during SLM, a melt pool is formed. "Shape" of this pool is a function of

https://doi.org/10.1515/9783110747232-005

a.  laser power (laser beam intensity) and
b.  presence of thermocapillary convection (Marangoni convection).

In even more advanced and recent models [885, 893], transfer of heat after its generation is considered by three main parameters:
a.  heat transfer due to convection,
b.  evaporation (i.e., formation of plasma; this results in re-radiation (inverse radiation)) and
c.  conduction from bottom and side walls.

This is a very recent and advanced approach which, however, ignores Marangoni convection effects. Overall, heat transfer phenomena associated with solidification of metal in liquid melt pool in AM are associated with three processes:
–  Generation of heat (laser matter interaction)
–  Assimilation of heat (melting and stages of solidification)
–  Extraction of heat

### 5.2.1 Generation of heat (laser matter interaction)

This is the first stage of AM in which heat is generated. The problem in this stage is related with impingement of light of certain intensity ($I$) on a solid surface for a certain amount of time which may result in production of heat. This interaction can be explained in terms of law known as "Beer Lambert's law."

#### 5.2.1.1 Beer Lambert's law for AM
Consider a thin layer of powder with thickness $d_1$ on a flat disk substrate of refractory metal with thickness $d_2$ and radius $r$ uniformly illuminated by light of intensity $I$.

For absorptivity of powder (or melt) assuming uniform temperature throughout the disk, the temperature evolution is

$$(\rho_1 c_1 d_1 + \rho_2 c_2 d_2)\frac{dT}{dt} = A(T)I - Q(T) \tag{11}$$

where $A(T)$ is absorptivity, $Q(T)$ is thermal loss (convective and radiative), $I$ is the intensity, $\rho_1$ is the density of powder, $\rho_2$ is the density of substrate, $c_1$ is the specific heat of powder, $c_2$ is the specific heat of substrate, $d_1$ is the thickness of powder and $d_2$ is the thickness of substrate.

Heat generated by this process is used for melt pool generation (its morphology, homogenization and holding (generation of supercooled liquid (SCL) region and its progression)).

### 5.2.1.2 Other recent approaches

Other recent approaches based on heat source, its nature and mode are described here. Essentially, these are based on how heat source models evolved from point source to modern-day complex three-dimensional models [1030–1032]. This evolution has been steady and progressive. It is described in detail here.

#### 5.2.1.2.1 Point and line heat source model

Point and line [1033, 1034] heat source models were first developed by Rosenthal. Essentially these were 1D models for moving heat source to analyze weld thermal field history which was an area of active research in the 1930s. However, as the name suggests, they were insufficient for more sophisticated problems involving heat distribution in two dimensions.

#### 5.2.1.2.2 Disk-shaped heat source model with Gaussian distribution of flux

To counter the above problems, a two-dimensional disk-shaped heat source model with Gaussian distribution of flux deposited on the surface of plate was developed [1035]. It was combined with the finite element method by Tekriwal et al. [1036] and Friedman [1037]. The model achieved more realistic temperature distribution than the previous Rosenthal model. However, it was unsuitable for high energy density welding such as laser welding. Another model was proposed.

#### 5.2.1.2.3 Volume source hemispherical power density distribution model

This model was proposed by Paley et al. [1038]. This model achieved reasonably accurate results for actual welding problems. It accounted for power density and its distribution into main Laplace equation and extended it to three dimensions which helped in simulating actual temperature, solute and density profiles around a moving heat source in a weld. This was a major achievement as it also accounted for and helped to understand energy distribution in complex high energy density laser welding processes as well. This model was further extended and finally evolved into double ellipsoid method explained below.

#### 5.2.1.2.4 Double-ellipsoid three-dimensional heat source model

As a result of requirement and progress in the field to go as near as possible to actual temperature and material profile during welding and laser processing, Goldak et al. improved upon the existing Paley model (Section 5.2.1.2.3) and proposed a new model [1039] known as double-ellipsoid three-dimensional heat source model. This was the major achievement of the time as it removed many discrepancies and bottlenecks in prior models. It was very true depiction of actual temperature profiles observed in complex multi-physics phenomena when a high energy density source such as laser strikes the sample.

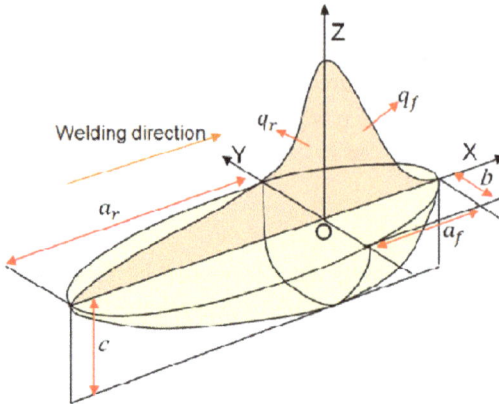

**Figure 35:** Double-ellipsoid heat source configuration along with power distribution function along $\xi$-axis.

### 5.2.1.2.5 Three-dimensional conical heat source model with interpolation

Though excellent, the Goldak heat source model was unsuitable to account for more complex multiphysics phenomena happening in fusion zone of weld metal pool and in cases where laser striking cause changes in grain structure. In such scenarios, interpolation in models is required. Recently Dezfoli et al. [1040] proposed a model which accounts for these factors. It also takes care of the type or mode of melt pool (conduction and keyhole), discontinuous melt track, formation of defects and pores employing not only primary laser source but secondary laser heat source. It employs interpolation wit an aim to decrease temperature gradient inside melt pool and shape of melt pool. Model is quite effective in retaining its accuracy and consistency.

### 5.2.2 Assimilation of heat (melting and stage of solidification)

As the heat generated above interacts with metal powder, it causes its melting and generation of liquid melt pool. The behavior of certain metal/alloy in melt pool can be explained by its cooling curve which is briefly described below.

### 5.2.2.1 General form of cooling curve

Cooling curve of a metal/alloy is a plot of variation of temperature with time. It has different regions which embody various types of information. Cooling curve can have different shapes depending on the metal or alloy type. A schematic cooling curve is shown in Figure 36 for a single-component pure metal (without any inoculants).

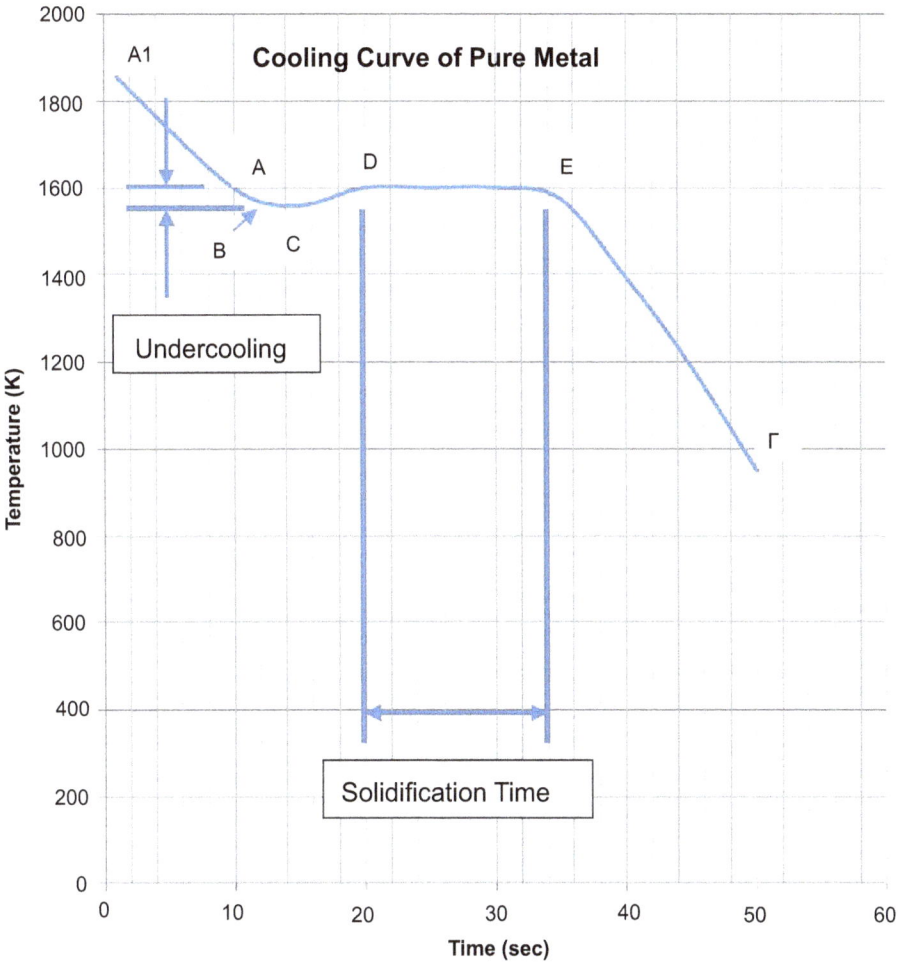

**Figure 36:** Cooling curve for a single-component pure metal (without any inoculants).

Its distinct regions are explained as follows:

Region above $A_1$: This is the region in which metal is in its complete liquid state and can be described by only melting and liquid state homogenization. Heat carried by metal in this region is "superheat" only and lost in the form of specific heat ($mc_p\Delta T$). This homogenization in turn depends on the type of melting (gas/solid (coal)-/liquid (oil)-fired crucible furnace melting, electric (resistance/induction/arc) melting) and subsequent melt treatment. (Note: Homogenization is required by some external means in the case of all modes of melting. Only induction furnace is manifested by self-homogenization due to phenomena of induction currents.)

Region $A_1$–A: This is the region which is characterized by the loss of superheat until the first arrest point A (the point of formation of first nucleant – explained in detail in later sections). This is also called start of solidification. In pure metals it is a sharp point (melting point), while in alloys, it can be a range (melting range). In BMGMCs/multicomponent alloys, it is also called start of supercooled region (SCL). This region is followed by undercooling ($\Delta T_n$) region which is described below.

Region A–D: This is the most important region of cooling curve (present case) for pure metals. In this region, metal cools down to a specific temperature characterized by a certain minimum amount of energy (activation energy for nucleation) needed to overcome a barrier of energy (energy barrier to nucleation) to create a liquid–solid (L–S) interface, eventually leading to formation of stable nuclei out of melt. This region is further divided into two regions: A–C and C–D.

Region A–C: This is the region in which undercooling occurs, heat is extracted, temperature drops and the shape of cooling curve goes down. This is characterized by two energies described in the above paragraph.

Region C–D: This is the region in which heat energy is absorbed, temperature is gained and shape of curve goes up. This is called recalescence.

Notes:
(a) Recalescence is gain in temperature as a result of thermal fluctuations caused by phase transformations occurring within solidifying melt/alloy.
(b) Region A–C is characterized by another point, Point "B" occurring in the middle of cooling curve. This is specifically shown in Figure 28 as an intermediate point of SCL. For the present case model (transient heat transfer conditions will be modeled at this point as well to get a better understanding of phenomena occurring in SCL in BMGMCs).

Region D–E: This is the region at which (after arrest point D) a metal losses all its heat of fusion ($mH_f$). In this region transformation occurs at a constant temperature in such a way that all liquid gets transformed into complete solid (all fine equiaxed grains formation at mold wall (Cu mold casting)/at surface of inoculant (heterogeneous nucleation – not present case), "equiaxed–columnar" transition, growth of columnar dendrites, CET (columnar-to-equiaxed transition) and growth of all equiaxed dendrites accomplishes). This is also called solidification time.

Region E–F: This is the region in which solid cools. That is, an alloy reaches it completion of solidification and cools to room temperature in solid state. This again occurs after a sharp invariant point (point F) in the case of pure metals and after a range in the case of multicomponent alloys.

### 5.2.2.2 Cooling curve for well-inoculated Zr-based in situ dendrite BMGMCs
The shape of cooling curve changes its form as melt is changed from a single to binary to multicomponent alloys. This can be explained in the form of various cases.

### Case I: Well-inoculated single-component melt

In these types of alloys, undercooling/undercooled region ($\Delta T_n$) diminishes and is almost absent. Inoculation with potent nuclei serves as active nucleation sites and triggers heterogeneous nucleation as the alloy reaches its first invariant point. Thus, no undercooling happens and solid alloy directly starts cooling as all liquid gets transformed to solid at constant temperature.

### Case II: Binary alloys without inoculants (slowly cooled)

In these types of alloys cooling occurs in the following steps:

1. Distinct undercooling occurs (characterized by drop and gain (recalescence) of temperature)
2. It is followed by a region of constant temperature cooling which is called *local solidification*. This is visible only in the case of very fluid alloys in which the mushy region is very fluid/less viscous (not BMGMCs). This region is absent in most multicomponent (industrial) alloys as their solidification is dominated by mushy zone. (Note: BMGMCs are special case of alloys in which the mushy region is extensively dominated but another phenomenon known as "sluggishness" governs the solidification. In these alloys, three laws [118] which describe BMGMC formation and evolution make sure that not only sluggishness dominates kinetics but it also ensures "glass formation" (i.e., retaining SCL at room temperature).
3. Alloy solidification range (it depends on the alloy). In slowly cooled binary alloys (most laboratory conditions), this is very clearly marked (usually bears an intermediate shape).
4. At the end of this range, alloy becomes stable momentarily at constant temperature (usually negligible in most industrial castings) at which nuclei (dendrite arm branches) grow and fills the interdendritic arm spacing and other small liquid pockets. This is marked by the end of solidification. (In some cases, it is also characterized by the start of CET and then growth of equiaxed grains.)
5. Following this point, solid alloy cools to room temperature or below room temperature (in the case of cryogenic cooling).

Note: For theoretical analysis, cooling curve can be of any type of combination between the type of alloy (single, binary and multicomponent), method of cooling (slow or fast) and inoculation (zero inoculation and well inoculated). All these can be drawn following rules of thermal transitions and kinetics. For simplicity and sufficiency, we will jump to cooling curve of multicomponent alloy (BMGMCs) fast cooled and well inoculated (present case).

**Case III: Multicomponent alloys with inoculants (fast cooled, in this case, BMGMCs)**
In these types of alloys, cooling can occur in the following steps (Figure 37):
1.  No undercooling occurs (as there is sufficient amount (number) of potent nuclei which serve as sites for active nucleation triggering heterogeneous nucleation prior to loss of temperature (drop of cooling curve), and gain of temperature (recalescence – rise of cooling curve)).

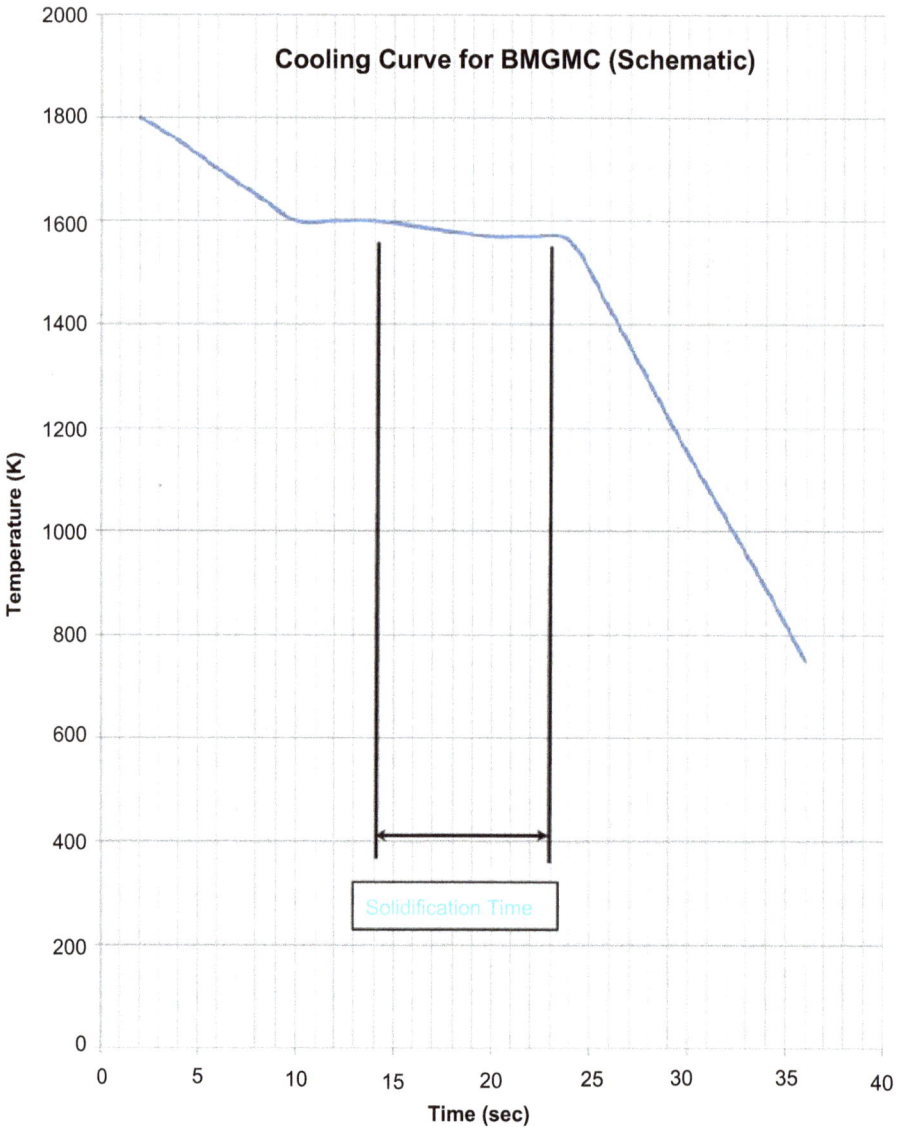

**Figure 37:** Cooling curve for multicomponent alloys with inoculants (fast-cooled BMGMC).

2. This is followed by a region of constant temperature at which all liquid gets transformed into solid. However, in these alloys, this region is very small (because of the presence of marked mushy zone).
3. Instantly after this region, alloy enters in alloy "solidification range." As the alloy is very fast cooled, this region is again not very clearly identified which is the typical behavior in the case of fast-cooled castings.
4. Following this, again alloy momentarily enters in brief constant temperature zone which marks the start of CET and growth of equiaxed grains (B2 CuZr phase equiaxed dendrites) until all liquid gets transformed into solid (end of solidification). This again is not very distinct as other phenomena (suppressing kinetics) dominate.
5. Finally, after this, BMGMC solidifies to room temperature.

Note: The shape of cooling curve in the case of slowly cooled and fast-cooled alloys is the slope of the curve toward the end of cooling which is very steep in the case of very fast-cooled alloys (liquid melt pools, in this case)).

### 5.2.2.3 Extraction of heat – determination of heat transfer coefficients

In the development of model, HTCs will be determined at every point of cooling curve following earlier defined 1D schemes [1041]. These will ensure the time of solidification calculation during cooling following the above cooling curve and helps in determining the shape of melt pool and its transient behavior during cooling.

### 5.2.2.4 Final time of solidification

The final time of solidification is the sum of time in each region/section of cooling curve of an alloy/melt. It will be determined using standard transport equations and will be used empirically to assess the conformability of AM process. The time of solidification gives other parameters as well such as fraction of mass solidified after a time $t$ which is a direct measure of microstructure evolved during that time. It can be qualitatively (extrapolation) used to predict further (type (equiaxed, columnar, mix and CET) and amount) evolution of microstructure with time.

## 5.3 Modeling and simulation of nucleation (heterogeneous) in liquid melt pool – microstructural development

Modeling and simulation of microstructural development in liquid melt pool can be described by macroscopic and microscopic models of heat and mass transfer depending on type of alloy, its nature, number of elements, cooling curve, undercoolings (constitutional (solute/particulate), curvature and interfacial), thermal and kinetic limitations, behavior of mushy zone, presence or absence of inoculants and so on.

These can be broadly divided into macroscopic and microscopic models [100] which are explained as follows.

### 5.3.1 Macroscopic models

By following the regimes of macroscopic models, finite element method (FEM) and finite difference method (FDM) can be used to explain microstructural development during both steady and transient state transport processes.

#### 5.3.1.1 Limitations

Both FEM- and FDM-based models cannot fully describe a mushy region, its behavior and evolution during solidification as they do not account for microscopic
- solute diffusion and
- capillary effects

which are primarily responsible for scale at which microstructure forms (which is very small as compared to macroscopic methods based on average continuity equations [1042–1045] in which it is assumed that solidification starts at liquidus and finishes at solidus/eutectic temperatures (a case of BMGMCs having good match of GFA and eutectic temperature [168, 169]). In order to overcome these limitations, microscopic models were proposed.

### 5.3.2 Microscopic models of microstructure evolution/formation during solidification

**Stage 1 model:** These models take into account the mechanism of (1) grain nucleation and (2) grain growth in alloys which are solidifying with equiaxed dendrite or eutectic microstructures [1046]. These do not account for alloys which are solidifying with columnar dendritic and planar interfaces. A modification of these accounts for equiaxed–columnar (at mold wall) and CET in bulk of liquid (this will be discussed later). These can be used to "describe microstructures" and "prediction of grain size" in the case of eutectic compositions of BMGMC. The majority of these are based on "analytical/deterministic approaches" which can be described as follows.

#### 5.3.2.1 Nucleation
- Choose a time "$t$" (initially nonzero value)
- At this time $t$, density of grains (which have nucleated in bulk) is a function of undercooling

$$d = f(\Delta T_n) \tag{12}$$

$f(\Delta T_n)$ is difficult to be found from theoretical considerations alone. It needs to be found experimentally, that is, from a set of experiments, for example:

*Method 1* Measurement of cooling curve.

This has been explained in detail in Sections 5.2.2.1 and 5.2.2.2.

*Method 2* Measurement of grain density (optical micrograph of cross section (using Image J®/manually)) for specimens solidified at various cooling rates [100].

### 5.3.2.2 Growth

As soon as grain has nucleated, and its growth can be explained by special modified case of CNT for BMGMC (a detailed treatment of modified CNT for BMGMC is given in Appendix A) and its distribution can be explained by constitutional supercooling zone/interdependence theory (propagation of L–S interface/L–S spherical front) (a possibility which is still under investigation by the author for suitability for AM processes), it grows with an interface velocity which is also a function of undercooling.

### 5.3.2.3 Velocity of growth

Velocity of growth may be written as

$$V_g = f\ (\Delta T_n) \tag{13}$$

In this case, there is no need to determine solidification kinetics of dendrite tip/eutectic (spherical front) interface by cooling curve or grain size, but it can be determined by theoretical models developed (by using basic laws of physics) [1047, 1048] as applied to BMGMC only under transient conditions.

### 5.3.2.4 Impingement

Impingement of grains as they grow is another important phenomenon which for all practical reasons governs the shape of grain after CET (CET in AM is recently explained by Amrita Basak et al. [1049] which is combined with the present model and is explained in detail in Appendix B) and is mainly responsible for equiaxed dendritic grain formation, especially in eutectic composition which is assumed to be the case for present research.

This has been typically treated by

a)  standard Johnson–Mehl–Avrami–Kolmogorov [1050, 1051] correction or by

b)  geometrical [1052, 1053] or

c)  random grain arrangement models [100].

These "microscopic" solidification models have been coupled with "macroscopic" transient 1D heat flow calculations to successfully predict "microstructural features" specially "grain size" at the scale of whole process (part scale) [1054, 1055].

### 5.3.2.5 Limitations

These deterministic models have their following limitations:

### a. Grain selection

They cannot account for the "grain selection" which occurs

a. close to mold region/surface giving rise to columnar dendritic microstructure (in the case of conventional Cu mold casting/TRC) or
b. at the surface of external inoculant particles (precursors of heterogeneous nucleation) in the case of well-inoculated melts (this case)) giving rise to onset of columnar dendritic microstructure (at a very small length scale) since they neglect any aspect which is related to crystallographic effects.

### b. "Equiaxed–columnar" transition

They cannot predict the so-called equiaxed–columnar transition which occurs very near the mold wall [1056] or variation of transverse size of columnar grains [1057] (also known as columnar dendritic arm branching). This is explained in detail in individual cases for each type of metal (crystal structure)

### i. Case 1: Cubic metals

It is a well-established fact that for cubic metals, this "grain selection" is based upon a criterion of "best alignment" of the <100> crystallographic axes of grain with heat flow direction [1056–1058]. Thus, this method cannot account for this anisotropic behavior of heat flow. A solution to this problem could be proposed by determining the best fit direction by use of recent developments in crystallography and their application to solidification. *Edge-to-edge matching* (*E2EM*): One way is to use the E2EM technique at the inoculant–ductile phase level (in the case of Zr-based BMGMC) (present research). This gives rise to selection of suitable potent nuclei of certain size and specific preferred orientation (i.e., along a defined easy crystallographic plane (e.g., (001)). If this crystallographic plane direction could be used in conjunction with macroscopic heat flow models, it can give rise to "prediction or selection of grain." In other words, if matching crystallographic axes (suitable for a potent inoculant selection for B2 ductile phase's preferred precipitation (in the case of BMGMC)) could be best aligned with heat flow direction (or heat flow direction could be assigned to this preferred matching crystallographic axes) a best "grain selection" could be determined (one of the aims of this research – not done previously elsewhere). This type of phenomena is particularly important in:

a.  Directional solidification or
b.  Production of single crystal dendritic alloys for aerospace applications or
c.  Production of BMGMC by Bridgeman solidification

Note: This is in addition to the use of E2EM for selection of potent nuclei.

### ii. Case 2: bcc metals

These methods are also ineffective in predicting "equiaxed–columnar" and then "branching of dendrite arms" in bcc metals (i.e., grain selection) as the best alignment between heat flow and crystallographic direction is not best known. Only assumptions are possible (i.e., in the case of bcc, the best heat flow direction could be assigned to close packed direction).

### iii. Case 3: fcc metals

These methods are again in effective in predicting the "equiaxed–columnar," "CET" and then branching of dendrite arms in fcc metals (i.e., grain selection) as the best alignment between heat flow and close packed direction (111) could only be assumed (to a satisfactory qualitative level). More quantitative experimentation is needed to determine best directions along which heat flow occurred or reverting to more advanced models.

### c. Extension of a grain into an open region of liquid

They cannot explain extension of a grain into an open region of liquid.

### d. Columnar-to-equiaxed transition

Finally, when very fine equiaxed grains at a region very close to mold wall/right at the interface of inoculant and melt are converted to columnar grains, which, when grow, there comes a point/plane at which columnar grains gets converted to not so fine equiaxed grains. This point is known as CET). These equiaxed grains finally extend toward the center of casting (wedge-shaped/melt pool centerline in the case of AM). CET primarily happens as a result of thermal fluctuations which happen at melt (liquid) and solid (solidified melt) interface which are triggered by solutal effects as well as heat extraction or absorption due to phase changes occurring at microscale (explained in subsequent sections). CET is dominant when thermal gradient is small.

### 5.3.3 Evolution of probabilistic models

The solution to the above four problems was presented first by Brown and Spittle [1059, 1060]. They developed probabilistic models. They used the Monte Carlo (MC) procedure for explaining solidification phenomena developed in earlier research [1061]. MC method is based upon minimizing of interfacial energy (which is practically calculated by using physical properties of material (Zr- and Fe-based BMGMC)) from literature and earlier published data or inference from extrapolation or interpolation of data as needed). Procedurally, this minimizations is achieved by

(a) Considering the energy of "unlike sites" (e.g., (a) "liquid/solid sites" or (b) "sites belonging to different grains" and
(b) By allowing transition between these states to occur according to randomly generated numbers

By using this method, Brown and Spittle were merely able to produce computed 2D microstructures which resembled very closely those observed in real micrographic cross section. In particular

a. The selection of grains in the columnar zone and
b. CET

were nicely reproduced using this technique. Also

a. the effect of solute concentration or
b. melt superheat upon the resultant microstructure

was determined "qualitatively" in a nice way. Their quantitative representation was not achieved.

### 5.3.3.1 Limitations

These methods suffer consistently from lack of physical basis and thus cannot be used to analyze quantitatively the effect of various physical phenomena (happening within the phase transformations). For example, to illustrate this, consider the following example:

a. During one MC time step, consider $N$ sites where $N$ is the number of sites whose evolution is calculated and is chosen from another $N$ (total number) sites. Therefore, not all sites of interest (i.e., those located near to solid–liquid interface) are investigated. This in turn, leads to algorithm predicted grain competition in columnar region, which does not at all reflect the physical mechanisms observed in organic alloys.
b. Furthermore, the results are sensitive to the type of MC network itself which is used for computations. Thus, a single powerful model is presented in this work which combines "advantages of probabilistic methods with those of deterministic approaches" to predict more accurately the grain structure in a casting.

### 5.3.4 Two-dimensional Cellular Automation (CA) method

For this purpose, for now, a 2D CA model is developed which is based upon physical mechanisms of nucleation and growth (NG) of dendritic grains. Its salient features are as follows.

1.  Heterogeneous Nucleation; which was modeled by means of a nucleation site distribution in deterministic solidification models, is treated in a similar way in present probabilistic approach.
2.  If the total density of grains which nucleate at a given undercooling is obtained from an average distribution ($d_c$ = average (distribution)), the location of these sites is chosen randomly

$$d_c = \frac{a_1 + a_2 + a_3 + a_4 + a_5 + \cdots + a_n}{n} \tag{14}$$

    where $a_1$, $a_2$, $a_3$, $a_4$,. . ., $a_n$ are distributions of grains 1, 2, 3 to $n$, $n$ = R (R = real numbers).
3.  Crystallographic orientation of a newly nucleated grain is also taken into account at random.
4.  The growth kinetics of (a) dendrite tip and (b) of side branches are also incorporated into the model in such a way that the final simulated microstructure is independent of the "CA network" which is used for computations.

Although it produces micrographic cross sections very much similar to those already obtained by Brown and Spittle, the present model has a "sound physical basis" and can thus reflect effects of (a) "cooling rate" or (b) "solute concentration" quantitatively.

### 5.3.4.1 Detailed description
Physical background: Consider a BMGMC wedge-shaped casting as shown in Figure 38.

Figure 38(c) is a typical 2D cross section of cast eutectic Zr-based BMGMC solidified in water-cooled wedge-shaped Cu mold [26, 1062]. Their dendritic grains which have various crystallographic orientations appear as zones of different colors (Figure 29(b)). Most common regions encountered in any casting appear here [1056, 1057] and are marked all along cross section. On the top end of the wedge-shaped ingot coarse grains are present as this region was exposed to air. Its more detailed explanation will follow after characterizing the region chronologically from bottom to top.

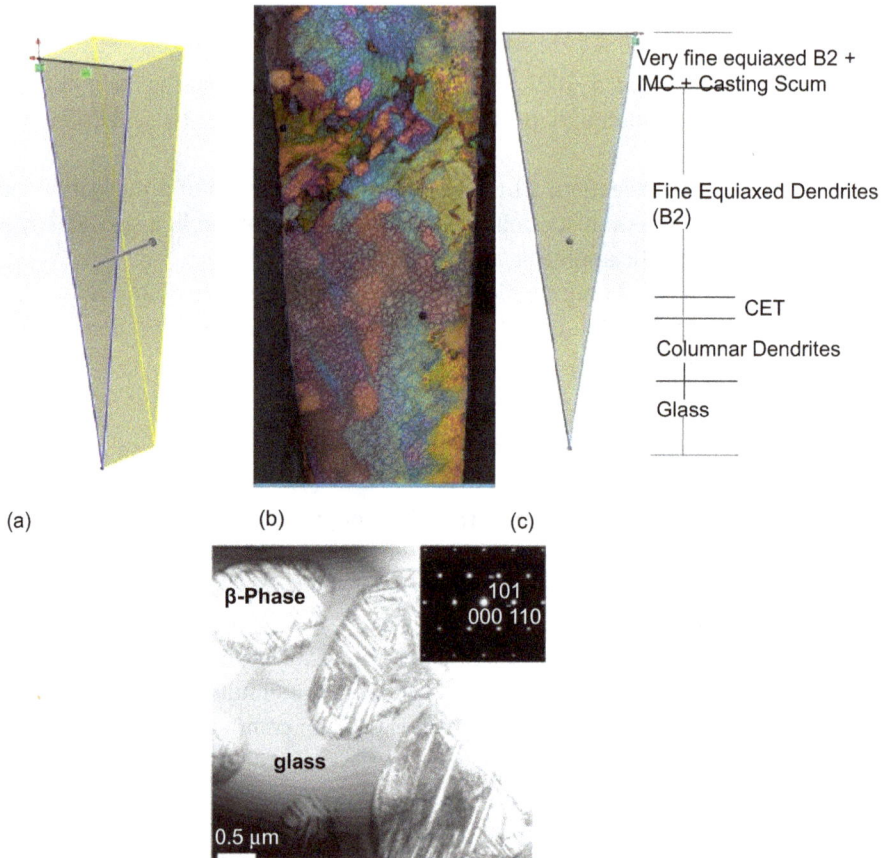

**Figure 38:** (a) Schematic 3Dm (b) optical micrograph of cross section (etched), (c) 2D schematic showing regions and (d) a specific region (from B2 dendrites) showing B19′ twins (B2–B19′ TRIP) [29].

### 5.3.4.2 Characterization

*Bottom region glass*: The tip of casting is 100% glass (monolithic BMG). This region is classified as glass and no crystal structure is observed here because cooling rate is maximum here which results in extraction of heat at a very high rate resulting in retaining an SCL state at room temperature.

*Bottom region columnar dendrites:* This region marks the beginning of "equiaxed–columnar" first transition. This consists of very fine layer in which this transition happens and then columnar grains grow (primarily) in random 3D orientation) because of still rapid rate of heat transfer which is complemented by a sluggish nucleation on growth mechanisms of BMGMC. These grains are not very long as the heat flow pattern is somewhat exponential because of the wedge-shaped casting which triggers the next transition too quickly before extension of growth as predicted by kinetics. This helps in retaining a glassy matrix all throughout the casting. Otherwise 100% crystallization would have happened.

*Bottom region CET:* This is the region in which columnar dendritic grains, which have developed/grown to a satisfactory level, transit to equiaxed grains, known as CET. This is triggered by various phenomena such as solute diffusion, solute–solvent partitioning, shape of liquid–solid propagation front and thermal fluctuations happening at the tip of L–S propagating interface.

*Fine equiaxed dendrites (B2):* Once CET happens, equiaxed dendrites are formed all throughout the casting. Only their shape differs. In this region, they are fine sized while in *Top Region*, their size is even more reduced due to presence of IMCs. Casting scum and other impurities combine with faster cooling rate from open top (convection and radiation) and side walls (conduction).

Note: In the case of BMGMC not only inoculant particles serve as sites for heterogeneous nucleation but grain boundaries also serve this purpose [346, 1063]. Other defects and solidification microstructure also serve as sites for heterogeneous nucleation. (Their effects in total solidification [nucleation and growth model] are to be taken into account in the final model.)

# Appendix A: Heterogeneous nucleation and growth in very fluid alloys (as per CNT) [1064]

Heterogeneous nucleation rate per unit volume is defined as

$$I = N_s \nu \exp\left(-\frac{\Delta G_d}{kT}\right) \exp\left(-\frac{\Delta G_c}{kT}\right) \tag{15}$$

where $N_s$ represents the number of atoms in contact with the substrate, $\nu$ is the vibrational frequency, $\Delta G_c$ is the activation energy for nucleation (critical energy of nucleus formation (i.e., creation of liquid–solid interface), $\Delta G_d$ is the activation energy of diffusion (diffusional activation energy).

Rearranging eq. (15) using definition of $\nu$ vibrational frequency

$$I = (N_0 - N)I_0 \exp\left(-\frac{\Delta G_c}{kT}\right)$$

$$I = (N_0 I_0 - N I_0) \exp\left(-\frac{\Delta G_c}{kT}\right) \tag{16}$$

*Proof.*

$$I = N_0 I_0 e^{\left(-\frac{\Delta G_c}{kT}\right)} - N I_0 e^{\left(-\frac{\Delta G_c}{kT}\right)} \tag{17}$$

$$I = N_0 \left(\frac{N_s}{t}\right) e^{\left(-\frac{\Delta G_c}{kT}\right)} - N \left(\frac{N_s}{t}\right) e^{\left(-\frac{\Delta G_c}{kT}\right)} \tag{18}$$

$$I = N_0 \left(\frac{N_s}{t}\right) e^{\left(-\frac{\Delta G_c}{kT}\right)} - N \left(\frac{N_s}{t}\right) e^{\left(-\frac{\Delta G_c}{kT}\right)} \tag{19}$$

$$I = (\nu \times t)\left(\frac{N_s}{t}\right) e^{\left(-\frac{\Delta G_c}{kT}\right)} - (\nu \times t)\left(\frac{N_s}{t}\right) e^{\left(-\frac{\Delta G_c}{kT}\right)} \tag{20}$$

$$I = (N_s \times \nu) e^{\left(-\frac{\Delta G_c}{kT}\right)} - (N \times \nu) e^{\left(-\frac{\Delta G_c}{kT}\right)} \tag{21}$$

$$I = \nu e^{\left(-\frac{\Delta G_c}{kT}\right)} (N_s - N) \tag{22}$$

$$I = \nu e^{\left(-\frac{\Delta G_c}{kT}\right)} N_s \tag{23}$$

$N$ can be neglected as during initial stages as there is no nucleation event.

According to classical nucleation theory (CNT), a minimum energy value is needed to create a solid–liquid interface eventually leading to stable nuclei out of melt. This is known as "activation energy." This activation energy is the energy to overcome $\Delta G^\star$ – the energy barrier to nucleation. Now, as solid–liquid interface grows to

https://doi.org/10.1515/9783110747232-006

form stable nuclei, atoms must be transported through liquid; thus, another temperature-dependent activation energy must be overcome known as $\Delta G_d$ (activation energy for diffusion).

The net effect is that CNT predicts a nucleation rate ($I$) given by

$$I = I_o \exp\left[-\frac{(\Delta G^* + \Delta G_d)}{k_\beta T}\right]$$

It is the nature of difference between $\Delta G^*$ and $\Delta G_d$ that dictates whether solidification will be crystalline or glassy. For crystalline solids, $\Delta G_d$ has a significant value while for glassy solids, there is no diffusion; thus, $\Delta G_d$ can be neglected. Thus

$$I = I_o \exp\left[-\frac{\Delta G^*}{k_\beta T}\right]$$

where $k_\beta$ is a constant dictated by the nature and type of liquid composition and measured experimentally. $\Delta G_d$ is also zero in case of small undercooling (i.e., well-inoculated liquids/multicomponent alloys (metallic glasses inoculated with/without potent nuclei (this study))) [520].

**Notes:**

1) **Vibrational frequency**

$$\frac{N_s}{t} = \upsilon = \frac{\text{occurrence of total number of heterogeneous substrate particles}}{\text{total time}}$$

$$\upsilon \times N_s$$

$$\frac{N_o}{t} \times N_s$$

where $N_s = I_o \times t$ or $I_o = (N_s/t)$, that is,

$$\text{Initial nucleation rate} = \frac{\text{total no. of atoms in contact with substrate}}{\text{total time}}$$

the definition used in eq. (17).

2) The difference between frequency and rate is that frequency is "occurrence of an event per unit time," while rate is the total number of that event (in terms of numerical value) per unit time.

Thus, from eq. (16), $N_o$ is the total number of heterogeneous substrate particles originally available per unit volume, $N$ is the number that has already nucleated and $I_o$ is constant.

Value of $I_0$ can be calculated from eq. (15) using another term known as *"liquid diffusion coefficient"*

$$D \approx a^2 \times \upsilon \exp\left(-\frac{\Delta G_d}{kT}\right) \tag{24}$$

where $a$ is the atomic diameter $= 0.4$ nm and $\upsilon$ is the frequency, which gives $I_0 = 10^{18} - 10^{22}/s$ for small values of undercooling (well-inoculated melts/multicomponent alloys)

$$\Delta G_c \propto 1/(\Delta T)^2$$

where $\Delta T$ represents undercooling.

Thus, nucleation rate is eq. (16)

$$I = \left[N_0 10^{20} - N10^{20}\right] \exp\left(-\frac{1}{kT(\Delta T)^2}\right)$$

or

$$I = \left[N_0 - N\right) 10^{20} \exp\left(-\frac{u}{(\Delta T)^2}\right) \tag{25}$$

where $u$ is a constant

$$u = \frac{1}{kT}$$

The value of $u$ can be measured from

Method 1: $T$ (heterogeneous nucleation temperature). This is defined as a temperature where there is an initial nucleation rate of one nucleus/cm$^3$/s.

Method 2: Second method to calculate $u$ is

$$u = -(\Delta T_N)^2 \ln\left(N_o \times 10^{20}\right)$$

*Proof.*

Taking natural log of eq. (25) on both sides,

$$\ln(I) = \ln(N_0 - N)10^{20}\left(-\frac{u}{(\Delta T_N)^2}\right)$$
$$-u = (\Delta T_N)^2\left[\ln I - \left[\ln(N_0 - N)10^{20}\right]\right]$$
$$-u = (\Delta T_N)^2\left[\ln I - \left[\ln N_0 10^{20} - \ln N10^{20}\right]\right]$$
$$-u = (\Delta T_N)^2\left[\ln I - \ln N_0 10^{20} + \ln N10^{20}\right]$$

because $\ln I$ and $\ln N10^{20}$ can be neglected

$$u = -\,(\Delta T_N)^2\ln I - \ln\left(N_0 \times 10^{20}\right)$$

where $\Delta T_N$ represents undercooling at heterogeneous nucleation temperature

Time is a user-defined input, and temperature comes from a user-defined value initially as well. Then its every new value is assigned back to eq. (15). With temperature and time, $k$ changes and assigned back to eq. (25). Also, with time, $v$ (vibrational frequency) changes and assigned back to the original equation (15). Similarly, the value of $u$ also changes with time and temperature. Table 5 summarizes the values that are user defined and that change as a function of transience as programs run.

**Table 5:** Summary of user-defined and program-determined functions used in CNT modified for BMGMC.

| S. no. | User-defined value | Time | Temp$_{(f)}$ |
|--------|--------------------|------|--------------|
| 1 | Time | | |
| 2 | Temp$_{(i)}$ | | |
| 3 | Temp$_{(f)}$ | ✓ | |
| 4 | K | ✓ | ✓ |
| 5 | Y | ✓ | ✓ |
| 6 | U | ✓ | ✓ |

**Note:**
1. In BMG (bulk metallic glasses), in some cases, due to slow motion of large atoms, only nucleation happens and growth never happens. In these cases, a new phenomenon known as soft impingement effects of crystals must be considered. These could be solutal/thermal. However, this is quite rare.
2. In general, in case of BMG, CNT cannot be applied alone to describe NG (nucleation and growth).
3. Constitutional supercooled zone (CSZ) and interdependence models cannot be applied because of very high ($\eta$) viscosity of BMG (and their sluggish nature). CSZ and interdependence theories are for less viscous/more fluid alloys.
4. A new concept, known as complex interdiffusion tensor [520], is much more helpful to describe NG in BMG.
5. Fick's law is not sufficient in its native form (i.e., linear form).

# Appendix B: Columnar to Equiaxed Transition (CET)

## 1 Special case of growth of "columnar microstructures"

The growth of columnar dendrites, which is initiated by nuclei that are formed at the mold interface (Cu mold casting/twin-roll casting of bulk metallic glass matrix composites (BMGMC); only if CSZ is suppressed – not the present case), is usually simulated in a much simpler way. Again, in this case, there is no need to use cooling curve measurements or grain size measurement but same growth kinetics models [7, 8] can be used to determine

    undercooling of eutectic front ($\Delta T_n$ eutectic) or
    undercooling of dendrite tips ($\Delta T_n$ dendrite tip) as well as
    undercooling of lamellae or dendrite trunk spacing ($\Delta T_n$ lamellae/trunk spacing).

This undercooling is determined by direct measurement of
(a) thermal gradient and
(b) speed of corresponding isotherm (eutectic or liquidus, respectively, that is, speed of eutectic isotherm and speed of liquidus isotherm).

The later values are obtained from a macroscopic (part scale) heat flow calculations [1041, 1046]. The secondary arm spacing of both equiaxed and columnar dendritic microstructures is deduced from a local solidification time.

## 2 Columnar structure growth in well-inoculated BMGMC

Growth of columnar dendrites can also occur at the surface of external inoculants (well-inoculated deeply undercooled melts – present case of BMGMC development). However, it should also be noted that another condition for growth of columnar dendrite to occur is suppression of CSZ which clashes with the aforementioned condition for onset of this phenomena at external potent nuclei of inoculant. That is why still there is dispute about the application of this concept to deeply undercooled well-inoculated melts (BMGMC) whose solution is under investigation.

https://doi.org/10.1515/9783110747232-007

## 3 Columnar to equiaxed transition (CET) [1049]

Growth rate of solid–liquid interface

$$V = S \cos \theta \tag{26}$$

where S is the scan speed.

Temperature gradient parallel to dendrite growth direction can be calculated using

$$G_{hkl} = G/\cos \psi \tag{27}$$

where $\psi$ is the angle between "normal vector" and "possible dendrite growth orientation" at the solid–liquid interface. This is evaluated by CFX–Post in Ansys®.

A modification known as Rappaz modification that is applied to predict columnar to equiaxed transition (CET) is given as follows:

$$\frac{G_{hkl}^n}{V_{hkl}} \geq a \left[ 3\sqrt{\frac{-4\pi}{3\ln(1-\varnothing)}} \sqrt{\frac{N_o}{n+1} \left( 1 - \frac{\Delta T_n^{n+1}}{\Delta T_{tip}^{n+1}} \right)} \right] \exp(n) \tag{28}$$

where $V_{hkl}$ is the dendrite growth velocity = $S \cos \theta/\cos \psi$, $n$ is the material constant determined from literature [82, 480], $\varphi$ is equiaxed fraction (critical value = 0.066%), $N_o$ is the nucleation density, $\Delta T_{tip}$ is tip undercooling and $\Delta T_n$ = nucleation undercooling.

This will be incorporated in the present model at point where CET is determined. However, this model does not give true 3D representation output.

Note: In general, phase field methods are for microstructure evolution (its type (planer front, spherical front, cellular) and morphology (precipitates, dendrites, platelets, acicular, needle like and spheroids)) while cellular automation methods are for grain size determination (equiaxed/columnar dendritic) and its prediction. If both are combined [975, 988, 1065], it is possible to get a full map of microstructure evolution and grain size.

# Comparison

A comparison of "strengths and capabilities" and "evolution of different theories over time," which enabled a better understanding of nucleation and growth phenomena in bulk metallic glass matrix composites (BMGMCs), is shown in Tables 6 and 7. The aim is to present a reader with a concise smart workable data for first-hand use and reference for solving nucleation and growth problems in BMGMCs by modeling and simulation. This will help professional programmer, working engineer and a researcher to effectively find previously done research till now with its strengths and capabilities at one platform.

## 8.1 Strengths and capabilities

A comparison of strength, capabilities and shortcomings of both deterministic and probabilistic methods is described in Table 6. It highlights and chalks out parameters and certain segments of each technique which could possibly advantageously used over others for modeling and simulation of BMGMCs.

https://doi.org/10.1515/9783110747232-008

**Table 6:** Comparison of strength and capabilities of modeling and simulation techniques as applied to nucleation and growth problem of bulk metallic glass matrix composites.

| S. no. | Phenomena/property | Deterministic models | | Probabilistic models | | Comments | References |
|---|---|---|---|---|---|---|---|
| | | Ductile phase | Glass | Ductile phase | Glass | | |
| 1 | Nucleation (heterogeneous) | ✓ | N/A | ✓ | N/A | | [1046] |
| 2 | Growth | ✓ | ✓ | × | × | | [520, 1064] |
| 3 | Growth mechanism (interdependence theory/complex interdiffusion tensor) | ✓ | N/A | ✓ | ✓ | | [520, 1066] |
| 4 | Different types of undercoolings (M-H model) | ✓ | ✓ | N/A | N/A | | [1067] |
| 5 | Growth kinetics | × | × | ✓ | ✓ | | [100] |
| 6 | Velocity of growth | ✓ | ✓ | ✓ | ✓ | | [100, 1047, 1048] |
| 7 | CET | ✓ | N/A | ✓ | N/A | Deterministic models can model ductile phase in 2D only | [100, 1049, 1068] |
| 8 | Impingement after CET | ✓ | N/A | ✓ | N/A | Deterministic models can model ductile phase by Johnson–Mehl–Avrami–Kolmogorov correction, geometrical and random grain arrangement models only | [100, 1050–1053] |

| No. | Feature | | | | Remarks | References |
|---|---|---|---|---|---|---|
| 9 | Grain selection — Qualitatively | × | ✓ | N/A | Probabilistic models can model ductile phase by MC only | [100] |
| | Grain selection — Quantitatively | | ✓ | | Probabilistic models can model ductile phase by CA only | |
| 10 | Columnar dendrite arm branching | × | × | ✓ | | [100, 104] |
| 11 | Extension of grain | × | × | ✓ | | [101, 104] |
| 12 | CET in 3D | × | N/A | ✓ | | [101, 103, 969] |
| 13 | Physical basis | N/A | N/A | ✓ | Probabilistic models can form the basis of modeling by MC; Probabilistic models form the basis of modeling by CA | [100, 104] |
| 14 | Quantitative | ✓ | ✓ | ✓ | Probabilistic models can only model quantitatively employing CA method | [100, 104] |
| 15 | Liquid–liquid transition | ✓ | N/A | N/A | | [91–93] |
| 16 | Devitrification | ✓ | ✓ | ✓ | Probabilistic models can model ductile and glass phases by 2D CA | [261, 1069] |

**Note:** N/A is an abbreviation to "not applicable." CET, columnar-to-equiaxed transition; CA, cellular automation; MC, Monte Carlo.

## 8.2 Evolution of theories

Table 7: Evolution of theories of modeling and simulation as applied to nucleation and growth problem of bulk metallic glass matrix composites.

| S. no. | Method/theory/approach | Action and explanation | Limitation to explanation | Group/institute | Year | Reference |
|---|---|---|---|---|---|---|
| | | **Part-scale modeling** | | | | |
| | | **Analytical modeling** | | | | |
| 1 | Deterministic/continuum model | Nonrandom methods produce same types of exact results | Does not depend on initial state/point | | 1993 | [967] |
| 2 | Probabilistic/stochastic models | Randomized result-based methods | Does depend on initial state | | 2016 | [961, 962] |
| | | **Computational modeling** | | | | |
| 3 | Lattice Boltzmann methods | Solution of basic "continuity" and "Navier–Stokes" equations for CFD based on Ludwig Boltzmann's kinetic theory of gases | Limited to CFD-type problems | Raabe, D. (MPIE, Dusseldorf) | 2004 | [977–983] |
| 4 | Phase field method | Solution of phase field parameter $\phi$ to describe physical state (liquid/solid) of material | Limited by type of $\phi$ for a particular situation | Napolitano, R. E. (Iowa State) | 2002 and 2012 | [94, 99] |
| 5 | Cellular automaton method | Division of entire volume into finite cells and solution of transport equations applied to individual cell | Large initial capital (processor/RAM) | Rappaz, M. (EPFL) | 1993 | [100, 101, 974, 989, 1070] |

| | | | | | |
|---|---|---|---|---|---|
| 6 | Virtual front tracking method | Dendritic growth in low Péclet number systems | Best in 2D | Stefanescu, D. M. (OSU) | 2007 | [990] |
| 7 | Sharp interface method | Evolution of interface as a function of time | Best in simple cases | Vermolen, F. J. (Delft) | 2006 | [996] |
| 8 | CAFE | Combine cellular automaton scheme with finite element method | | Rappaz, M. (EPFL) | 1994 | [104] |
| 9 | PFFE | Combine phase field with finite element method | | Britta Nestler (KIT) | 2011 | [987] |
| 10 | PFCA | Combine phase field with cellular automata regime | | Shin, Y. C. (Purdue) | 2011 | [1065] |
| **Atomistic Modeling** | | | | | |
| 10 | Classical MD | Exact solutions | Computing power | Alder and Wainwright (Lawrence Livermore) | 1957 | [158, 696, 949] |
| 11 | Monte Carlo simulation | Set of probability-based possible outcomes | Range of solutions | Metropolis, Nicholas and coworkers (LANL) | 1953 and 1993 | [574, 1005, 1006, 1008] |
| 12 | Ab initio method/first principle calculation | Based on solution of Schrödinger equation | Works well for H atom only. For all other atoms, approximations are needed | Robert Parr (Caltech) | 1950 | [1013] |
| 13 | Hartree Fork method and Slater determinant | Uses the variational theorem (which is a wave function-based approach using mean field approximation) | Approximate solution is obtained. It is a form of ab initio method | | | [1012] |

(continued)

Table 7 (continued)

| S. no. | Method/theory/approach | Action and explanation | Limitation to explanation | Group/institute | Year | Reference |
|---|---|---|---|---|---|---|
| 14 | Evolution of Hartree Fork method | | | | | |
| | Self-constrained field method | Evolution of Hartree Fork method | Approximate solutions | | | [1012] |
| | Møller–Plesset (MP) perturbation (MP 1) | Hamiltonian is divided into two parts $\hat{H} = \hat{H}_0 + \lambda\hat{V}$ and solved | $\psi$ and energy are HF $\psi$ and HF energy | | | [1012] |
| | MP 2 | $\psi$ remains the same and energy is changed | $\psi$ is treated by the help of summations | | | [1012] |
| | Density functional theory | Energy of system is obtained from electron density | Approximation based | | 1996 | [1071–1073] |
| 15 | Interatomic potential | Explain interaction of atoms in a system in terms of potentials | Limited by accuracy, transferability and computational speed of system | Multi (Many) Body Potentials, Daw Baska (Sandia National Labs) | 1984 | [1021] |

# Research gap

Nucleation and growth (NG) phenomena in single (pure metals), binary and multicomponent alloys are rather well understood. Classical nucleation theory (CNT) [1050] provides many answers to behavior of these melts. Bulk metallic glass (BMG) and their composites (BMGMCs) are relatively new class of materials that have recently emerged on the surface of science and technology and gained attention due to their unique properties [10, 139, 228, 1074]. Traditionally, they were produced using conventional methods (Cu mold casting [352, 354, 1075] and twin-roll casting (TRC) [350]) in which their metastable phase (glass) and any in situ ductile precipitates (stable phase) are nucleated based on their ability to surpass activation energy barrier. In addition, these processes impart very high cooling rate to castings, which are essential for retention of supercooled liquid (glass) at room temperature explained by phenomena of confusion [127], ordering [126, 1076, 1077], frustration [125] and vitrification [1078, 1079].

Very recently, with the advent and popularity of additive manufacturing (AM), interest has sparked to exploit the inherent and fundamental advantages present in this unique process to produce BMG and BMGMCs. AM techniques are useful in achieving this objective as very high cooling rate in fusion liquid melt pool is already present inherently to assist the formation of glassy structure which is suppression of "kinetics" and prolonging of undercooling ("thermodynamics") – two main phenomena responsible for any phase transformation. However, the in situ nucleation of second phase equiaxed dendrites during solidification and then microstructural evolution (*solute diffusion* and *capillary* assisted) are not satisfactorily explained by CNT alone.

Either some modifications are needed in CNT or more reliable probabilistic microstructure evolution models (e.g., Johnson–Mehl–Avrami–Kolmogorov correction [520]) are needed to explain NG (and other phenomena, e.g., liquid–liquid transition [91, 92, 695] and phase separations [257]) in BMGMCs. In this work, an effort has been made to meet both requirements, and propositions are given as follows:

At present scenario, there is no single hybrid/combined model which explains phenomena of heat transfer (liquid melt pool formation as a result of laser–matter interaction and its evolution–solidification), and coupled this with NG (solute diffusion [1080] and capillary action driven) at microscale to predict microstructure and grain size in BMGMC as melt cools in liquid pool of AM.

A. Only one study has been conducted to model the same phenomena (solidification only) during Cu mold suction casting which will serve as base [1081] in addition to very recent attempts [520] in which emphasis is laid on development of generalized theory rather than solving a problem.

B. Only one study has been reported on microstructure formation during TRC using CAFE [351] but that is not aimed at BMGMCs is carried out using commercial

https://doi.org/10.1515/9783110747232-009

software package and does not involve any mathematical modeling at the back end. Software embedded (NG and heat transfer) models are only used.

C. Four prominent studies, namely, by Zhou et al. [111], Zhang et al. [112], Zinoviev et al. [109] and a group at Shenyang, China [110, 973], have been reported very recently using CAFE but these are based on modeling microstructure evolution in modified AM hybrid deposition micro rolling [HDMR] [111], LAMP [75] on 316L SS [112], 2D CAFE] [109], and cladding [110, 973]) processes.

D. Few studies in the past have been conducted employing SLM using CAFE [110, 973, 1082, 1083], cellular automaton phase field (CAPF), cellular automaton finite volume method (CAFVM) [1084], modified CAFE [1085] and other approaches but none have been conducted on BMGMC.

E. No effort has been made to correlate the effect of edge to edge method (E2EM) with the assigning direction of easy heat flow and easy crystallographic growth.

F. No substantial study has been reported about evolution of microstructure in three dimensions in BMGMCs in AM.

G. No effort has been made to combine the effect of changing properties with a decrease in temperature (transient conditions). Most of the models till now predict solutions in terms of steady-state processes.

H. Very few studies have been carried out to combine cellular automation with finite element in case of AM while it is a routine approach to predict grain size in case of other processes (casting and welding).

# Overall aims/research questions

Following research questions will be addressed in this study:

1. How can potent inoculants be designed for controlling number density ($d_c$) and distribution of ductile phase in bulk metallic glass matrix composites (BMGMC) via solidification?
2. What is the effect of number density ($d_c$), size and distribution of ductile phase on final mechanical properties of BMGMCs?
3. What is the link between crystallographic matching model and heat flow in case of Zr-based BMGMCs?
4. For laser surface remelting process, what is the role of changing process variables (transient conditions – thermal conductivity of mold, temperature of melt, temperature of mold, viscosity of melt, etc.) on final microstructure evolution of BMGMC?
5. How do multiple thermal cycles affect the final microstructure of BMGMC processed by additive manufacturing (AM)?
6. How does coupling of modeling and simulation of microstructural evolution and heat transfer in BMGMC during AM using advanced programming and simulation platforms help in understanding nucleation and growth phenomena?

https://doi.org/10.1515/9783110747232-010

# Methodology

Following methodology will be adopted to realize the conceived model and its simulation:

**Step 1:** Write codes for transient conditions in MatLab using standard transport equations.

- Heat transfer equations to calculate time during each step
- Mass transfer (nucleation and growth dominated by solute diffusion [1080] and capillary) equations to determine grain size during solidification steps only

**Step 2:** Make part/section drawings representing liquid melt pool in additive manufacturing (AM) in Ansys®/SolidWorks®

    Import/integrate MatLab code
    Determine mesh size and optimize it
    Select properties (from literature and software libraries) and surfaces
    Run simulation

The following scheme will be used to make simulation sets. A total of 125 simulation sets were run, whose results will be followed in subsequent publications.

| Sr. No | No. Density (dc) | Particle Size (micron) | Distribution |
|---|---|---|---|
| 1 | 10 | 5 | 50 |
| 2 | | | 100 |
| 3 | | | 150 |
| 4 | | | 200 |
| 5 | | | 250 |

| Sr. No | No. Density (dc) | Particle Size (micron) | Distribution |
|---|---|---|---|
| 1 | 20 | 5 | 50 |
| 2 | | | 100 |
| 3 | | | 150 |
| 4 | | | 200 |
| 5 | | | 250 |

| Sr. No | No. Density (dc) | Particle Size | Distribution |
|---|---|---|---|
| 1 | 10 | 10 | 50 |
| 2 | | | 100 |
| 3 | | | 150 |
| 4 | | | 200 |
| 5 | | | 250 |

| Sr. No | No. Density (dc) | Particle Size | Distribution |
|---|---|---|---|
| 1 | 20 | 10 | 50 |
| 2 | | | 100 |
| 3 | | | 150 |
| 4 | | | 200 |
| 5 | | | 250 |

| Sr. No | No. Density (dc) | Particle Size | Distribution |
|---|---|---|---|
| 1 | 10 | 15 | 50 |
| 2 | | | 100 |
| 3 | | | 150 |
| 4 | | | 200 |
| 5 | | | 250 |

| Sr. No | No. Density (dc) | Particle Size | Distribution |
|---|---|---|---|
| 1 | 20 | 15 | 50 |
| 2 | | | 100 |
| 3 | | | 150 |
| 4 | | | 200 |
| 5 | | | 250 |

| Sr. No | No. Density (dc) | Particle Size | Distribution |
|---|---|---|---|
| 1 | 10 | 20 | 50 |
| 2 | | | 100 |
| 3 | | | 150 |
| 4 | | | 200 |
| 5 | | | 250 |

| Sr. No | No. Density (dc) | Particle Size | Distribution |
|---|---|---|---|
| 1 | 20 | 20 | 50 |
| 2 | | | 100 |
| 3 | | | 150 |
| 4 | | | 200 |
| 5 | | | 250 |

| Sr. No | No. Density (dc) | Particle Size | Distribution |
|---|---|---|---|
| 1 | 10 | 25 | 50 |
| 2 | | | 100 |
| 3 | | | 150 |
| 4 | | | 200 |
| 5 | | | 250 |

| Sr. No | No. Density (dc) | Particle Size | Distribution |
|---|---|---|---|
| 1 | 20 | 25 | 50 |
| 2 | | | 100 |
| 3 | | | 150 |
| 4 | | | 200 |
| 5 | | | 250 |

Total Number of Samples        125

https://doi.org/10.1515/9783110747232-011

| Sr. No | No. Density (dc) | Particle Size (micron) | Distribution |
|---|---|---|---|
| 1 | 30 | 5 | 50 |
| 2 | | | 100 |
| 3 | | | 150 |
| 4 | | | 200 |
| 5 | | | 250 |

| Sr. No | No. Density (dc) | Particle Size | Distribution |
|---|---|---|---|
| 1 | 30 | 10 | 50 |
| 2 | | | 100 |
| 3 | | | 150 |
| 4 | | | 200 |
| 5 | | | 250 |

| Sr. No | No. Density (dc) | Particle Size | Distribution |
|---|---|---|---|
| 1 | 30 | 15 | 50 |
| 2 | | | 100 |
| 3 | | | 150 |
| 4 | | | 200 |
| 5 | | | 250 |

| Sr. No | No. Density (dc) | Particle Size | Distribution |
|---|---|---|---|
| 1 | 30 | 20 | 50 |
| 2 | | | 100 |
| 3 | | | 150 |
| 4 | | | 200 |
| 5 | | | 250 |

| Sr. No | No. Density (dc) | Particle Size | Distribution |
|---|---|---|---|
| 1 | 30 | 25 | 50 |
| 2 | | | 100 |
| 3 | | | 150 |
| 4 | | | 200 |
| 5 | | | 250 |

| Sr. No | No. Density (dc) | Particle Size (micron) | Distribution |
|---|---|---|---|
| 1 | 40 | 5 | 50 |
| 2 | | | 100 |
| 3 | | | 150 |
| 4 | | | 200 |
| 5 | | | 250 |

| Sr. No | No. Density (dc) | Particle Size (micron) | Distribution |
|---|---|---|---|
| 1 | 40 | 10 | 50 |
| 2 | | | 100 |
| 3 | | | 150 |
| 4 | | | 200 |
| 5 | | | 250 |

| Sr. No | No. Density (dc) | Particle Size (micron) | Distribution |
|---|---|---|---|
| 1 | 40 | 15 | 50 |
| 2 | | | 100 |
| 3 | | | 150 |
| 4 | | | 200 |
| 5 | | | 250 |

| Sr. No | No. Density (dc) | Particle Size (micron) | Distribution |
|---|---|---|---|
| 1 | 40 | 20 | 50 |
| 2 | | | 100 |
| 3 | | | 150 |
| 4 | | | 200 |
| 5 | | | 250 |

| Sr. No | No. Density (dc) | Particle Size (micron) | Distribution |
|---|---|---|---|
| 1 | 40 | 25 | 50 |
| 2 | | | 100 |
| 3 | | | 150 |
| 4 | | | 200 |
| 5 | | | 250 |

| Sr. No | No. Density (dc) | Particle Size (micron) | Distribution |
|---|---|---|---|
| 1 | 50 | 5 | 50 |
| 2 | | | 100 |
| 3 | | | 150 |
| 4 | | | 200 |
| 5 | | | 250 |

| Sr. No | No. Density (dc) | Particle Size (micron) | Distribution |
|---|---|---|---|
| 1 | 50 | 10 | 50 |
| 2 | | | 100 |
| 3 | | | 150 |
| 4 | | | 200 |
| 5 | | | 250 |

| Sr. No | No. Density (dc) | Particle Size (micron) | Distribution |
|---|---|---|---|
| 1 | 50 | 15 | 50 |
| 2 | | | 100 |
| 3 | | | 150 |
| 4 | | | 200 |
| 5 | | | 250 |

| Sr. No | No. Density (dc) | Particle Size (micron) | Distribution |
|---|---|---|---|
| 1 | 50 | 20 | 50 |
| 2 | | | 100 |
| 3 | | | 150 |
| 4 | | | 200 |
| 5 | | | 250 |

| Sr. No | No. Density (dc) | Particle Size (micron) | Distribution |
|---|---|---|---|
| 1 | 50 | 25 | 50 |
| 2 | | | 100 |
| 3 | | | 150 |
| 4 | | | 200 |
| 5 | | | 250 |

**Step 3: Experimental validation of microstructures**
For verification of simulation results of microstructural development (modeling), the following observation/experimental approach will be adopted.

**Observation "in situ"**
   High-speed camera to observe melt pool in AM and/or
   Radiation pyrometer with imaging system and/or
   Thermocouples (temperature measurement and its correlation with microstructure; different phases solidify at different cooling rates, which can give information about the type of microstructure evolved during solidification)
   Optical micrography of actual castings (Cu mold wedge)/AM part will be carried out after appropriate etching (type, number density, size and distribution will tally with simulated results and correlations drawn)

**Determination of number density ($d_c$) from TTT diagram**
Get time–temperature transformation (TTT) diagram of exact alloy composition (bulk metallic glass matrix composites, BMGMC) [165, 174] and determine the critical cooling rate just enough to form (nucleate) a certain percentage of one crystalline phase in another (in present case, glass). Then design apparatus accordingly. The objective is not to form a monolithic glass but a crystal phase embedded glass (i.e., ductile glass). Further, TTT diagram is used to calculate the volume fraction of phases anticipated and these are compared with simulated and experimental results.

## 11.1 Progress so far

Ductile BMGMC samples have been casted in wedge shape using Cu mold suction casting at CSIRO, Calyton, Vic. Two main compositions, namely, eutectic ($Zr_{47}Cu_{45.5}Al_5Co_2$) and other hypoeutectic $Zr_{65}Cu_{15}Al_{10}Ni_{10}$ (described in detail in Section 2.1.13) are casted.
1.  $Zr_{47}Cu_{45.5}Al_5Co_2$: Optical (Figure 31) and scanning electron microscopies (backscatter electron imaging (Figure 34)) have been performed on $Zr_{47}Cu_{45.5}Al_5Co_2$ after etching with solution of HF (hydrofluoric acid) and $HNO_3$ in water.
2.  Scanning electron microscopy (secondary electron imaging (Figure 32)) has been performed on $Zr_{65}Cu_{15}Al_{10}Ni_{10}$ after light etching with $HNO_3$.

The micrographs taken are shown in the following figures and they clearly indicate the presence of ductile B2–CuZr phase in the form of spheroids originating from liquid upon solidification (Figures 39 and 40). Furthermore, etched surfaces of BMGMCs show pitting (Figures 41 and 42), which is a characteristic feature observed in this class of alloys upon treatment with HF acid.

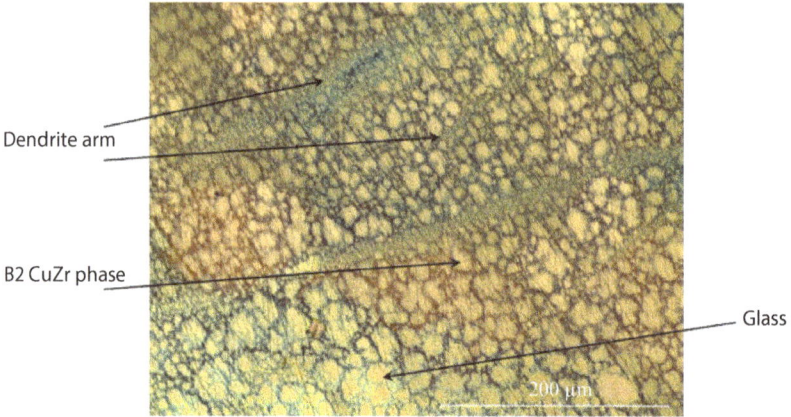

**Figure 39:** Incident light optical micrographs (without polarizer) of cross section of a wedge sample of $Zr_{47}Cu_{45.5}Al_5Co_2$ (etched: HF + $HNO_3$ + $H_2O$).

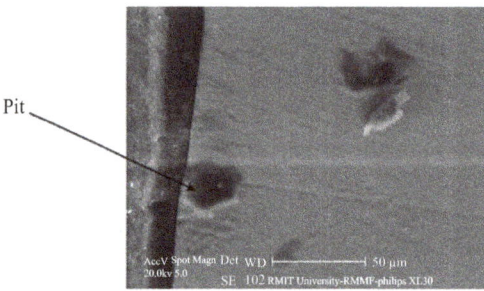

**Figure 40:** Secondary electron image of cross section of a wedge sample of $Zr_{65}Cu_{15}Al_{10}Ni_{10}$ (light etch: $HNO_3$ + $H_2O$).

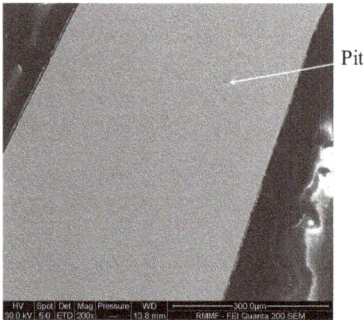

**Figure 41:** Low magnification secondary electron (SE) image of cross section of a wedge sample of $Zr_{47}Cu_{45.5}Al_5Co_2$ (etched: HF + $HNO_3$ + $H_2O$) (showing pits) in glassy matrix).

These features, their appearance and effect of number density, size and distribution of ductile phase will be studied with modeling and simulation during the upcoming years to develop a deeper and better understanding of nucleation phenomena in BMGMCs.

Glassy
Matrix

B2 CuZr

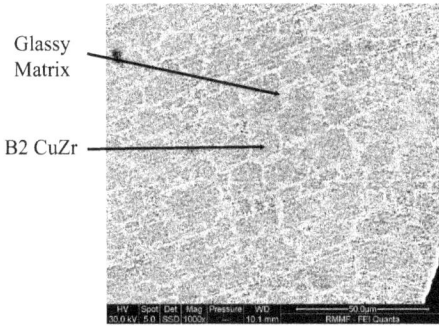

**Figure 42:** Backscatter electron (BSE) image of cross section of a wedge sample of $Zr_{47}Cu_{45.5}Al_5Co_2$ (etched: HF + $HNO_3$ + $H_2O$) (clearly showing B2–CuZr phase (spheroids), interdendritic network and glassy matrix).

# References

[1]     Klement, W., R.H. Willens and P.O.L. Duwez, Non-crystalline structure in solidified gold-silicon alloys. Nature, 1960. 187(4740): 869–870.

[2]     Telford, M., The case for bulk metallic glass. Materials Today, 2004. 7(3): 36–43.

[3]     Schuh, C.A., T.C. Hufnagel and U. Ramamurty, Mechanical behavior of amorphous alloys. Acta Materialia, 2007. 55(12): 4067–4109.

[4]     Inoue, A. and A. Takeuchi, Recent development and application products of bulk glassy alloys. Acta Materialia, 2011. 59(6): 2243–2267.

[5]     Chen, H.S., Thermodynamic considerations on the formation and stability of metallic glasses. Acta Metallurgica, 1974. 22(12): 1505–1511.

[6]     Drehman, A.J., A.L. Greer and D. Turnbull, Bulk formation of a metallic glass: Pd40Ni40P20. Applied Physics Letters, 1982. 41(8): 716–717.

[7]     Kui, H.W., A.L. Greer and D. Turnbull, Formation of bulk metallic glass by fluxing. Applied Physics Letters, 1984. 45(6): 615–616.

[8]     Wang, W.H., C. Dong and C.H. Shek, Bulk metallic glasses. Materials Science and Engineering: R: Reports, 2004. 44(2–3): 45–89.

[9]     Cheng, Y.Q. and E. Ma, Atomic-level structure and structure–property relationship in metallic glasses. Progress in Materials Science, 2011. 56(4): 379–473.

[10]    Qiao, J., H. Jia and P.K. Liaw, Metallic glass matrix composites. Materials Science and Engineering: R: Reports, 2016. 100: 1–69.

[11]    Trexler, M.M. and N.N. Thadhani, Mechanical properties of bulk metallic glasses. Progress in Materials Science, 2010. 55(8): 759–839.

[12]    Hays, C.C., C.P. Kim and W.L. Johnson, Microstructure controlled shear band pattern formation and enhanced plasticity of bulk metallic glasses containing formed ductile phase dendrite dispersions. Physical Review Letters, 2000. 84(13): 2901–2904.

[13]    Hofmann, D.C., et al., Designing metallic glass matrix composites with high toughness and tensile ductility. Nature, 2008. 451(7182): 1085–1089.

[14]    Hofmann, D.C., Shape Memory bulk metallic glass composites. Science, 2010. 329(5997): 1294–1295.

[15]    Wu, Y., et al., Designing bulk metallic glass composites with enhanced formability and plasticity. Journal of Materials Science & Technology, 2014. 30(6): 566–575.

[16]    Guo, H., et al., Tensile ductility and necking of metallic glass. Nat Mater, 2007. 6(10): 735–739.

[17]    Jang, D. and J.R. Greer, Transition from a strong-yet-brittle to a stronger-and-ductile state by size reduction of metallic glasses. Nat Mater, 2010. 9(3): 215–219.

[18]    Choi-Yim, H., Synthesis and Characterization of Bulk Metallic Glass Matrix Composites. 1998, California Institute of Technology.

[19]    Choi-Yim, H., et al., Processing, microstructure and properties of ductile metal particulate reinforced Zr57Nb5Al10Cu15.4Ni12.6 bulk metallic glass composites. Acta Materialia, 2002. 50(10): 2737–2745.

[20]    Lee, M.L., Y. Li and C.A. Schuh, Effect of a controlled volume fraction of dendritic phases on tensile and compressive ductility in La-based metallic glass matrix composites. Acta Materialia, 2004. 52(14): 4121–4131.

[21]    Pauly, S., et al., Transformation-mediated ductility in CuZr-based bulk metallic glasses. Nat Mater, 2010. 9(6): 473–477.

[22]    Wu, Y., et al., Bulk metallic glass composites with transformation-mediated work-hardening and ductility. Advanced Materials, 2010. 22(25): 2770–2773.

https://doi.org/10.1515/9783110747232-012

[23]   Song, K.K., et al., Triple yielding and deformation mechanisms in metastable Cu47.5Zr47.5Al5 composites. Acta Materialia, 2012. 60(17): 6000–6012.

[24]   Wu, D.Y., et al., Glass-forming ability, thermal stability of B2 CuZr phase, and crystallization kinetics for rapidly solidified Cu–Zr–Zn alloys. Journal of Alloys and Compounds, 2016. 664: 99–108.

[25]   Kim, C.P., et al., Realization of high tensile ductility in a bulk metallic glass composite by the utilization of deformation-induced martensitic transformation. Scripta Materialia, 2011. 65(4): 304–307.

[26]   Gao, W.-H., et al., Effects of Co and Al addition on martensitic transformation and microstructure in ZrCu-based shape memory alloys. Transactions of Nonferrous Metals Society of China, 2015. 25(3): 850–855.

[27]   Zhai, H., H. Wang and F. Liu, A strategy for designing bulk metallic glass composites with excellent work-hardening and large tensile ductility. Journal of Alloys and Compounds, 2016. 685: 322–330.

[28]   Song, W., et al., Microstructural control via copious nucleation manipulated by in situ formed nucleants: Large-sized and ductile metallic glass composites. Advanced Materials, 2016. n/a–n/a.

[29]   Pekarskaya, E., C.P. Kim and W.L. Johnson, In situ transmission electron microscopy studies of shear bands in a bulk metallic glass based composite. Journal of Materials Research, 2001. 16(09): 2513–2518.

[30]   Zhang, Q., H. Zhang, Z. Zhu and H. Zhuangqi, Formation of High strength in-situ bulk metallic glass composite with enhanced plasticity in Cu50Zr47:5Ti2:5 Alloy. Materials Transactions, 2005. 46(3): 730–733.

[31]   Zhu, Z., et al., Ta-particulate reinforced Zr-based bulk metallic glass matrix composite with tensile plasticity. Scripta Materialia, 2010. 62(5): 278–281.

[32]   Fan, C., R.T. Ott and T.C. Hufnagel, Metallic glass matrix composite with precipitated ductile reinforcement. Applied Physics Letters, 2002. 81(6): 1020–1022.

[33]   Hu, X., et al., Glass forming ability and in-situ composite formation in Pd-based bulk metallic glasses. Acta Materialia, 2003. 51(2): 561–572.

[34]   Cheng, J.-L., et al., Innovative approach to the design of low-cost Zr-based BMG composites with good glass formation. Scientific Reports, 2013. 3: 2097.

[35]   Wu, F.F., et al., Effect of annealing on the mechanical properties and fracture mechanisms of a Zr 56.2 Ti 13.8 Nb 5.0 Cu 6.9 Ni 5.6 Be 12.5 bulk-metallic-glass composite. Physical Review B, 2007. 75(13): 134201.

[36]   Chen, H.S., Ductile-brittle transition in metallic glasses. Materials Science and Engineering, 1976. 26(1): 79–82.

[37]   Antonione, C., et al., Phase separation in multicomponent amorphous alloys. Journal of Non-Crystalline Solids, 1998. 232–234: 127–132.

[38]   Fan, C., C. Li and A. Inoue, Nanocrystal composites in Zr–Nb–Cu–Al metallic glasses. Journal of Non-Crystalline Solids, 2000. 270(1–3): 28–33.

[39]   Fan, C. and A. Inoue, Ductility of bulk nanocrystalline composites and metallic glasses at room temperature. Applied Physics Letters, 2000. 77(1): 46–48.

[40]   Basu, J., et al., Microstructure and mechanical properties of a partially crystallized La-based bulk metallic glass. Philosophical Magazine, 2003. 83(15): 1747–1760.

[41]   Fan, C., et al., Properties of as-cast and structurally relaxed Zr-based bulk metallic glasses. Journal of Non-Crystalline Solids, 2006. 352(2): 174–179.

[42]   Gu, J., et al., Effects of annealing on the hardness and elastic modulus of a Cu36Zr48Al8Ag8 bulk metallic glass. Materials & Design, 2013. 47: 706–710.

[43]  Tan, J., et al., Correlation Between Internal States and Strength in Bulk Metallic Glass, in PRICM. 2013, John Wiley & Sons, Inc. 3199–3206.

[44]  Krämer, L., et al., Production of bulk metallic glasses by severe plastic deformation. Metals, 2015. 5(2): 720.

[45]  Nishiyama, N., et al., The world's biggest glassy alloy ever made. Intermetallics, 2012. 30: 19–24.

[46]  Inoue, A., N. Nishiyama and T. Matsuda, Preparation of bulk glassy $Pd_{40}Ni_{10}Cu_{30}P_{20}$ alloy of 40 mm in diameter by water quenching. Materials Transactions, JIM, 1996. 37(2): 181–184.

[47]  He, Y., R.B. Schwarz and J.I. Archuleta, Bulk glass formation in the Pd–Ni–P system. Applied Physics Letters, 1996. 69(13): 1861–1863.

[48]  Inoue, A., T. Zhang and T. Masumoto, Zr–Al–Ni amorphous alloys with high glass transition temperature and significant supercooled liquid region. Materials Transactions, JIM, 1990. 31(3): 177–183.

[49]  Peker, A. and W.L. Johnson, A highly processable metallic glass: Zr41.2Ti13.8Cu12.5Ni10.0Be22.5. Applied Physics Letters, 1993. 63(17): 2342–2344.

[50]  Tan, J., et al., Study of mechanical property and crystallization of a ZrCoAl bulk metallic glass. Intermetallics, 2011. 19(4): 567–571.

[51]  Biffi, C.A., A. Figini Albisetti and A. Tuissi, CuZr based shape memory alloys: effect of Cr and Co on the martensitic transformation, In: Materials Science Forum. 2013, Trans Tech Publ.

[52]  Cheng, J.L. and G. Chen, Glass formation of Zr–Cu–Ni–Al bulk metallic glasses correlated with L → Zr2Cu + ZrCu pseudo binary eutectic reaction. Journal of Alloys and Compounds, 2013. 577: 451–455.

[53]  Chen, G., et al., Enhanced plasticity in a Zr-based bulk metallic glass composite with in situ formed intermetallic phases. Applied Physics Letters, 2009. 95(8): 081908.

[54]  Jeon, C., et al., Effects of effective dendrite size on tensile deformation behavior in Ti-based dendrite-containing amorphous matrix composites modified from Ti-6Al-4V alloy. Metallurgical and Materials Transactions A, 2015. 46(1): 235–250.

[55]  Chu, M.Y., et al., Quasi-static and dynamic deformation behaviors of an in-situ Ti-based metallic glass matrix composite. Journal of Alloys and Compounds, 2015. 640: 305–310.

[56]  Wang, Y.S., et al., The role of the interface in a Ti-based metallic glass matrix composite with in situ dendrite reinforcement. Surface and Interface Analysis, 2014. 46(5): 293–296.

[57]  Gibson, I., W.D. Rosen and B. Stucker, Development of Additive Manufacturing Technology, In: Additive Manufacturing Technologies: Rapid Prototyping to Direct Digital Manufacturing. 2010, Springer US:Boston, MA. 36–58.

[58]  Spears, T.G. and S.A. Gold, In-process sensing in selective laser melting (SLM) additive manufacturing. Integrating Materials and Manufacturing Innovation, 2016. 5(1): 1–25.

[59]  Pauly, S., et al., Processing metallic glasses by selective laser melting. Materials Today, 2013. 16(1–2): 37–41.

[60]  Schroers, J., Processing of Bulk Metallic Glass. Advanced Materials, 2010. 22(14): 1566–1597.

[61]  Li, X.P., et al., Selective laser melting of Zr-based bulk metallic glasses: Processing, microstructure and mechanical properties. Materials & Design, 2016. 112: 217–226.

[62]  Zheng, B., et al., Processing and Behavior of Fe-Based Metallic Glass Components via Laser-Engineered Net Shaping. Metallurgical and Materials Transactions A, 2009. 40(5): 1235–1245.

[63]  Olakanmi, E.O., R.F. Cochrane and K.W. Dalgarno, A review on selective laser sintering/melting (SLS/SLM) of aluminium alloy powders: Processing, microstructure, and properties. Progress in Materials Science, 2015. 74: 401–477.

[64]    Buchbinder, D., et al., High Power Selective Laser Melting (HP SLM) of Aluminum Parts. Physics Procedia, 2011. 12: 271–278.

[65]    Li, Y. and D. Gu, Thermal behavior during selective laser melting of commercially pure titanium powder: Numerical simulation and experimental study. Additive Manufacturing, 2014. 1–4: 99–109.

[66]    Yap, C.Y., et al., Review of selective laser melting: Materials and applications. Applied Physics Reviews, 2015. 2(4): 041101.

[67]    Romano, J., et al., Temperature distribution and melt geometry in laser and electron-beam melting processes – A comparison among common materials. Additive Manufacturing, 2015. 8: 1–11.

[68]    Sun, H. and K.M. Flores, Microstructural Analysis of a Laser-Processed Zr-Based Bulk Metallic Glass. Metallurgical and Materials Transactions A, 2010. 41(7): 1752–1757.

[69]    Yang, G., et al., Laser solid forming Zr-based bulk metallic glass. Intermetallics, 2012. 22: 110–115.

[70]    Zhang, Y., et al., Microstructural analysis of Zr55Cu30Al10Ni5 bulk metallic glasses by laser surface remelting and laser solid forming. Intermetallics, 2015. 66: 22–30.

[71]    Frazier, W.E., Metal additive manufacturing: a review. Journal of Materials Engineering and Performance, 2014. 23(6): 1917–1928.

[72]    Wong, K.V. and A. Hernandez, A review of additive manufacturing. ISRN Mechanical Engineering, 2012. 2012: 10.

[73]    Travitzky, N., et al., Additive manufacturing of ceramic-based materials. Advanced Engineering Materials, 2014. 16(6): 729–754.

[74]    Baufeld, B., E. Brandl and O. Van Der Biest, Wire based additive layer manufacturing: Comparison of microstructure and mechanical properties of Ti–6Al–4V components fabricated by laser-beam deposition and shaped metal deposition. Journal of Materials Processing Technology, 2011. 211(6): 1146–1158.

[75]    Chen, Y., C. Zhou and J. Lao, A layerless additive manufacturing process based on CNC accumulation. Rapid Prototyping Journal, 2011. 17(3): 218–227.

[76]    Kawahito, Y., et al., High-power fiber laser welding and its application to metallic glass Zr55Al10Ni5Cu30. Materials Science and Engineering: B, 2008. 148(1–3): 105–109.

[77]    Kim, J.H., et al., Pulsed Nd:YAG laser welding of Cu54Ni6Zr22Ti18 bulk metallic glass. Materials Science and Engineering: A, 2007. 449–451: 872–875.

[78]    Li, B., et al., Laser welding of Zr45Cu48Al7 bulk glassy alloy. Journal of Alloys and Compounds, 2006. 413(1–2): 118–121.

[79]    Wang, G., et al., Laser welding of Ti40Zr25Ni3Cu12Be20 bulk metallic glass. Materials Science and Engineering: A, 2012. 541: 33–37.

[80]    Wang, H.S., et al., Combination of a Nd:YAG laser and a liquid cooling device to (Zr53Cu30Ni9Al8)Si0.5 bulk metallic glass welding. Materials Science and Engineering: A, 2010. 528(1): 338–341.

[81]    Wang, H.-S., et al., The effects of initial welding temperature and welding parameters on the crystallization behaviors of laser spot welded Zr-based bulk metallic glass. Materials Chemistry and Physics, 2011. 129(1–2): 547–552.

[82]    Acharya, R. and S. Das, Additive manufacturing of IN100 superalloy through scanning laser epitaxy for turbine engine hot-section component repair: process development, modeling, microstructural characterization and process control. Metallurgical and Materials Transactions A, 2015. 46(9): 3864–3875.

[83]    Sames, W.J., et al., The metallurgy and processing science of metal additive manufacturing. International Materials Reviews, 2016. 61(5): 315–360.

[84]    Harooni, A., et al., Processing window development for laser cladding of zirconium on
        zirconium alloy. Journal of Materials Processing Technology, 2016. 230: 263–271.
[85]    Wu, X. and Y. Hong, Fe-based thick amorphous-alloy coating by laser cladding. Surface and
        Coatings Technology, 2001. 141(2–3): 141–144.
[86]    Wu, X., B. Xu and Y. Hong, Synthesis of thick Ni66Cr5Mo4Zr6P15B4 amorphous alloy
        coating and large glass-forming ability by laser cladding. Materials Letters, 2002. 56(5):
        838–841.
[87]    Yue, T.M. and Y.P. Su, Laser cladding of SiC reinforced Zr65Al7.5Ni10Cu17.5 amorphous
        coating on magnesium substrate. Applied Surface Science, 2008. 255(5): Part 1) 1692–1698.
[88]    Yue, T.M., Y.P. Su and H.O. Yang, Laser cladding of Zr65Al7.5Ni10Cu17.5 amorphous alloy
        on magnesium. Materials Letters, 2007. 61(1): 209–212.
[89]    Zhang, P., et al., Synthesis of Fe–Ni–B–Si–Nb amorphous and crystalline composite
        coatings by laser cladding and remelting. Surface and Coatings Technology, 2011. 206(6):
        1229–1236.
[90]    Zhu, Q., et al., Synthesis of Fe-based amorphous composite coatings with low purity
        materials by laser cladding. Applied Surface Science, 2007. 253(17): 7060–7064.
[91]    Lan, S., et al., Structural crossover in a supercooled metallic liquid and the link to a liquid-
        to-liquid phase transition. Applied Physics Letters, 2016. 108(21): 211907.
[92]    Zu, F.-Q., Temperature-induced liquid-liquid transition in metallic melts: a brief review on
        the new physical phenomenon. Metals, 2015. 5(1): 395.
[93]    Wei, S., et al., Liquid–liquid transition in a strong bulk metallic glass-forming liquid. Nature
        Communications, 2013. 4.
[94]    Boettinger, W.J., et al., Phase-field simulation of solidification. Annual Review of Materials
        Research, 2002. 32(1): 163–194.
[95]    Emmerich, H., Phase-field modelling for metals and colloids and nucleation therein – an
        overview. Journal of Physics: Condensed Matter, 2009. 21(46): 464103.
[96]    Emmerich, H., et al., Phase-field-crystal models for condensed matter dynamics on atomic
        length and diffusive time scales: an overview. Advances in Physics, 2012. 61(6): 665–743.
[97]    Gong, X. and K. Chou, Phase-field modeling of microstructure evolution in electron beam
        additive manufacturing. JOM, 2015. 67(5): 1176–1182.
[98]    Gránásy, L., et al., Phase-field modeling of polycrystalline solidification: from needle
        crystals to spherulites – a review. Metallurgical and Materials Transactions A, 2014.
        45(4): 1694–1719.
[99]    Wang, T. and R.E. Napolitano, A phase-field model for phase transformations in glass-
        forming alloys. Metallurgical and Materials Transactions A, 2012. 43(8): 2662–2668.
[100]   Rappaz, M. and C.A. Gandin, Probabilistic modelling of microstructure formation in
        solidification processes. Acta Metallurgica et Materialia, 1993. 41(2): 345–360.
[101]   Charbon, C. and M. Rappaz, 3D probabilistic modelling of equiaxed eutectic solidification.
        Modelling and Simulation in Materials Science and Engineering, 1993. 1(4): 455.
[102]   Gandin, C.A., R.J. Schaefer and M. Rappax, Analytical and numerical predictions of
        dendritic grain envelopes. Acta Materialia, 1996. 44(8): 3339–3347.
[103]   Gandin, C.A. and M. Rappaz, A 3D cellular automaton algorithm for the prediction of
        dendritic grain growth. Acta Materialia, 1997. 45(5): 2187–2195.
[104]   Gandin, C.A. and M. Rappaz, A coupled finite element-cellular automaton model for the
        prediction of dendritic grain structures in solidification processes. Acta Metallurgica et
        Materialia, 1994. 42(7): 2233–2246.
[105]   Chen, S., G. Guillemot and C.-A. Gandin, 3D coupled cellular automaton (CA)–finite element
        (FE) modeling for solidification grain structures in gas tungsten arc welding (GTAW).
        ISIJ International, 2014. 54(2): 401–407.

[106]   Chen, S., Three Dimensional Cellular Automaton-Finite Element (CAFE) Modeling for the Grain Structures Development in Gas Tungsten/Metal Arc Welding Processes. 2014, Ecole Nationale Supérieure des Mines de Paris.

[107]   Tsai, D.-C. and W.-S. Hwang, A three dimensional cellular automaton model for the prediction of solidification morphologies of brass alloy by horizontal continuous casting and its experimental verification. Materials Transactions, 2011. 52(4): 787–794.

[108]   Wei, L., et al., Low artificial anisotropy cellular automaton model and its applications to the cell-to-dendrite transition in directional solidification. Materials Discovery.

[109]   Zinoviev, A., et al., Evolution of grain structure during laser additive manufacturing. Simulation by a cellular automata method. Materials & Design, 2016. 106: 321–329.

[110]   Wang, Z.-J., et al., Simulation of microstructure during laser rapid forming solidification based on cellular automaton. Mathematical Problems in Engineering, 2014. 2014: 9.

[111]   Zhou, X., et al., Simulation of microstructure evolution during hybrid deposition and micro-rolling process. Journal of Materials Science, 2016. 51(14): 6735–6749.

[112]   Zhang, J., et al. Probabilistic simulation of solidification microstructure evolution during laser-based metal deposition. In Proceedings of 2013 Annual International Solid Freeform Fabrication Symposium–An Additive Manufacturing Conference. 2013.

[113]   Greer, A.L., Metallic Glasses. Science, 1995. 267(5206): 1947–1953.

[114]   Güntherodt, H.J., Metallic glasses, In: Festkörperprobleme 17: Plenary Lectures of the Divisions "Semiconductor Physics" "Metal Physics" "Low Temperature Physics" "Thermodynamics and Statistical Physics" "Crystallography" "Magnetism" "Surface Physics" of the German Physical Society Münster, March 7–12, 1977, Treusch, J., Editor. 1977, Springer Berlin Heidelberg:Berlin, Heidelberg. 25–53.

[115]   Inoue, A., High strength bulk amorphous alloys with low critical cooling rates (overview). Materials Transactions, JIM, 1995. 36(7): 866–875.

[116]   Johnson, W.L., Bulk glass-forming metallic alloys: science and technology. MRS Bulletin, 1999. 24(10): 42–56.

[117]   Matthieu, M., Relaxation and physical aging in network glasses: a review. Reports on Progress in Physics, 2016. 79(6): 066504.

[118]   Hofmann, D.C. and W.L. Johnson, Improving Ductility in Nanostructured Materials and Metallic Glasses:"Three Laws", in Materials Science Forum. 2010, Trans Tech Publ.

[119]   Shi, Y. and M.L. Falk, Does metallic glass have a backbone? The role of percolating short range order in strength and failure. Scripta Materialia, 2006. 54(3): 381–386.

[120]   Mattern, N., et al., Short-range order of Cu–Zr metallic glasses. Journal of Alloys and Compounds, 2009. 485(1–2): 163–169.

[121]   Jiang, M.Q. and L.H. Dai, Short-range-order effects on intrinsic plasticity of metallic glasses. Philosophical Magazine Letters, 2010. 90(4): 269–277.

[122]   Zhang, F., et al., Composition-dependent stability of the medium-range order responsible for metallic glass formation. Acta Materialia, 2014. 81: 337–344.

[123]   Sheng, H.W., et al., Atomic packing and short-to-medium-range order in metallic glasses. Nature, 2006. 439(7075): 419–425.

[124]   Cheng, Y.Q., E. Ma and H.W. Sheng, Atomic level structure in multicomponent bulk metallic glass. Physical Review Letters, 2009. 102(24): 245501.

[125]   Nelson, D.R., Order, frustration, and defects in liquids and glasses. Physical Review B, 1983. 28(10): 5515–5535.

[126]   Ma, E., Tuning order in disorder. Nat Mater, 2015. 14(6): 547–552.

[127]   Greer, A.L., Confusion by design. Nature, 1993. 366(6453): 303–304.

[128]   Chen, H.S., Glassy metals. Reports on Progress in Physics, 1980. 43(4): 353.

[129]   Turnbull, D., Under what conditions can a glass be formed?. Contemporary Physics, 1969. 10(5): 473–488.

[130]   Akhtar, D., B. Cantor and R.W. Cahn, Diffusion rates of metals in a NiZr2 metallic glass. Scripta Metallurgica, 1982. 16(4): 417–420.

[131]   Akhtar, D., B. Cantor and R.W. Cahn, Measurements of diffusion rates of Au in metal-metal and metal-metalloid glasses. Acta Metallurgica, 1982. 30(8): 1571–1577.

[132]   Akhtar, D. and R.D.K. Misra, Impurity diffusion in a Ni–Nb metallic glass. Scripta Metallurgica, 1985. 19(5): 603–607.

[133]   Inoue, A., T. Zhang and T. Masumoto, Glass-forming ability of alloys. Journal of Non-Crystalline Solids, 1993. 156: 473–480.

[134]   Lu, Z.P., Y. Liu and C.T. Liu, Evaluation of Glass-Forming Ability, In: Bulk Metallic Glasses, Miller, M. and P. Liaw, Editors. 2008, Springer US: Boston, MA. 87–115.

[135]   Yi, J., et al., Glass-forming ability and crystallization behavior of Al86Ni9La5 Metallic glass with Si addition. Advanced Engineering Materials, 2016. 18(6): 972–977.

[136]   Wang, L.-M., et al., A "universal" criterion for metallic glass formation. Applied Physics Letters, 2012. 100(26): 261913.

[137]   Donald, I.W. and H.A. Davies, Prediction of glass-forming ability for metallic systems. Journal of Non-Crystalline Solids, 1978. 30(1): 77–85.

[138]   Park, E.S. and D.H. Kim, Design of bulk metallic glasses with high glass forming ability and enhancement of plasticity in metallic glass matrix composites: A review. Metals and Materials International, 2005. 11(1): 19–27.

[139]   Chen, M., A brief overview of bulk metallic glasses. NPG Asia Mater, 2011. 3: 82–90.

[140]   Park, E.S., H.J. Chang and D.H. Kim, Effect of addition of Be on glass-forming ability, plasticity and structural change in Cu–Zr bulk metallic glasses. Acta Materialia, 2008. 56(13): 3120–3131.

[141]   Guo, G.-Q., S.-Y. Wu and L. Yang, Structural origin of the enhanced glass-forming ability induced by microalloying Y in the ZrCuAl alloy. Metals, 2016. 6(4): 67.

[142]   Cheng, Y.Q., E. Ma and H.W. Sheng, Alloying strongly influences the structure, dynamics, and glass forming ability of metallic supercooled liquids. Applied Physics Letters, 2008. 93(11): 111913.

[143]   Jia, P., et al., A new Cu–Hf–Al ternary bulk metallic glass with high glass forming ability and ductility. Scripta Materialia, 2006. 54(12): 2165–2168.

[144]   Miracle, D.B., et al., An assessment of binary metallic glasses: correlations between structure, glass forming ability and stability. International Materials Reviews, 2010. 55(4): 218–256.

[145]   Lu, Z.P. and C.T. Liu, A new glass-forming ability criterion for bulk metallic glasses. Acta Materialia, 2002. 50(13): 3501–3512.

[146]   Lu, Z.P. and C.T. Liu, A new approach to understanding and measuring glass formation in bulk amorphous materials. Intermetallics, 2004. 12(10–11): 1035–1043.

[147]   Li, Y., et al., Glass forming ability of bulk glass forming alloys. Scripta Materialia, 1997. 36(7): 783–787.

[148]   Kim, Y.C., et al., Glass forming ability and crystallization behavior of Ti-based amorphous alloys with high specific strength. Journal of Non-Crystalline Solids, 2003. 325(1–3): 242–250.

[149]   Wu, J., et al., New insight on glass-forming ability and designing Cu-based bulk metallic glasses: The solidification range perspective. Materials & Design, 2014. 61: 199–202.

[150]   Shen, T.D., B.R. Sun and S.W. Xin, Effects of metalloids on the thermal stability and glass forming ability of bulk ferromagnetic metallic glasses. Journal of Alloys and Compounds, 2015. 631: 60–66.

[151]   Li, P., et al., Glass forming ability, thermodynamics and mechanical properties of novel Ti–Cu–Ni–Zr–Hf bulk metallic glasses. Materials & Design, 2014. 53: 145–151.

[152]   Li, P., et al., Glass forming ability and thermodynamics of new Ti-Cu-Ni-Zr bulk metallic glasses. Journal of Non-Crystalline Solids, 2012. 358(23): 3200–3204.

[153]   Li, F., et al., Structural origin underlying poor glass forming ability of Al metallic glass. Journal of Applied Physics, 2011. 110(1): 013519.

[154]   Fan, C., et al., Effects of Nb addition on icosahedral quasicrystalline phase formation and glass-forming ability of Zr–Ni–Cu–Al metallic glasses. Applied Physics Letters, 2001. 79(7): 1024–1026.

[155]   Yang, H., K.Y. Lim and Y. Li, Multiple maxima in glass-forming ability in Al–Zr–Ni system. Journal of Alloys and Compounds, 2010. 489(1): 183–187.

[156]   Xu, D., G. Duan and W.L. Johnson, Unusual glass-forming ability of bulk amorphous alloys based on ordinary metal copper. Physical Review Letters, 2004. 92(24): 245504.

[157]   Zhang, K., et al., Computational studies of the glass-forming ability of model bulk metallic glasses. The Journal of Chemical Physics, 2013. 139(12): 124503.

[158]   Amokrane, S., A. Ayadim and L. Levrel, Structure of the glass-forming metallic liquids by ab-initio and classical molecular dynamics, a case study: quenching the Cu60Ti20Zr20 alloy. Journal of Applied Physics, 2015. 118(19): 194903.

[159]   Inoue, A. and A. Takeuchi, Bulk amorphous, nano-crystalline and nano-quasicrystalline alloys IV. Recent progress in bulk glassy alloys. Materials Transactions, 2002. 43(8): 1892–1906.

[160]   Inoue, A., B. Shen and A. Takeuchi, Developments and applications of bulk glassy alloys in late transition metal base system. Materials Transactions, 2006. 47(5): 1275–1285.

[161]   Weinberg, M.C., et al., Critical cooling rate calculations for glass formation. Journal of Non-Crystalline Solids, 1990. 123(1): 90–96.

[162]   Ray, C.S., et al., A new DTA method for measuring critical cooling rate for glass formation. Journal of Non-Crystalline Solids, 2005. 351(16–17): 1350–1358.

[163]   Kim, J.-H., et al., Estimation of critical cooling rates for glass formation in bulk metallic glasses through non-isothermal thermal analysis. Metals and Materials International, 2005. 11(1): 1–9.

[164]   Zhu, D.M., et al., Method for estimating the critical cooling rate for glass formation from isothermal TTT data, In: Key Engineering Materials. 2007, Trans Tech Publ.

[165]   Weinberg, M.C., D.R. Uhlmann and E.D. Zanotto, "Nose method" of calculating critical cooling rates for glass formation. Journal of the American Ceramic Society, 1989. 72(11): 2054–2058.

[166]   Inoue, A., Stabilization of metallic supercooled liquid and bulk amorphous alloys. Acta Materialia, 2000. 48(1): 279–306.

[167]   Wang, D., et al., Bulk metallic glass formation in the binary Cu–Zr system. Applied Physics Letters, 2004. 84(20): 4029–4031.

[168]   Lee, D.M., et al., A deep eutectic point in quaternary Zr–Ti–Ni–Cu system and bulk metallic glass formation near the eutectic point. Intermetallics, 2012. 21(1): 67–74.

[169]   Ma, D., et al., Correlation between glass formation and type of eutectic coupled zone in eutectic alloys. Materials Transactions, 2003. 44(10): 2007–2010.

[170]   Jian, X. Complete composition tunability of Cu (Ni)-Ti-Zr alloys for bulk metallic glass formation.

[171]   Ma, H., et al., Doubling the critical size for bulk metallic glass formation in the Mg–Cu–Y ternary system. Journal of Materials Research, 2011. 20(9): 2252–2255.

[172]   Lu, Z.P. and C.T. Liu, Glass formation criterion for various glass-forming systems. Physical Review Letters, 2003. 91(11): 115505.

[173]  Lad, K.N., Correlation between atomic-level structure, packing efficiency and glass-forming ability in Cu–Zr metallic glasses. Journal of Non-Crystalline Solids, 2014. 404: 55–60.

[174]  Mukherjee, S., et al., Overheating threshold and its effect on time–temperature-transformation diagrams of zirconium based bulk metallic glasses. Applied Physics Letters, 2004. 84(24): 5010–5012.

[175]  Brazhkin, V.V., Metastable phases and 'metastable' phase diagrams. Journal of Physics: Condensed Matter, 2006. 18(42): 9643.

[176]  Baricco, M., et al., Metastable phases and phase diagrams. La Metallurgia Italiana, 2004(11).

[177]  Brazhkin, V.V., Metastable phases, phase transformations, and phase diagrams in physics and chemistry. Physics-Uspekhi, 2006. 49(7): 719–724.

[178]  Taub, A.I. and F. Spaepen, The kinetics of structural relaxation of a metallic glass. Acta Metallurgica, 1980. 28(12): 1781–1788.

[179]  Tsao, S.S. and F. Spaepen, Structural relaxation of a metallic glass near equilibrium. Acta Metallurgica, 1985. 33(5): 881–889.

[180]  Akhtar, D. and R.D.K. Misra, Effect of thermal relaxation on diffusion in a metallic glass. Scripta Metallurgica, 1986. 20(5): 627–631.

[181]  Qiao, J.C. and J.M. Pelletier, Dynamic mechanical relaxation in bulk metallic glasses: a review. Journal of Materials Science & Technology, 2014. 30(6): 523–545.

[182]  Liu, C., E. Pineda and D. Crespo, Mechanical relaxation of metallic glasses: an overview of experimental data and theoretical models. Metals, 2015. 5(2): 1073.

[183]  Wen, P., et al., Mechanical relaxation in supercooled liquids of bulk metallic glasses. Physica Status Solidi (a), 2010. 207(12): 2693–2703.

[184]  Levine, D. and P.J. Steinhardt, Proceedings of the international conference on the theory of the structures of non-crystalline solids quasicrystals. Journal of Non-Crystalline Solids, 1985. 75(1): 85–89.

[185]  Steinhardt, P.J., Quasicrystals: a new form of matter. Endeavour, 1990. 14(3): 112–116.

[186]  Janot, C., The structure of quasicrystals. Journal of Non-Crystalline Solids, 1993. 156: 852–864.

[187]  Xing, L.Q., et al., Effect of cooling rate on the precipitation of quasicrystals from the Zr–Cu–Al–Ni–Ti amorphous alloy. Applied Physics Letters, 1998. 73(15): 2110–2112.

[188]  Xing, L.Q., et al., High-strength materials produced by precipitation of icosahedral quasicrystals in bulk Zr–Ti–Cu–Ni–Al amorphous alloys. Applied Physics Letters, 1999. 74(5): 664–666.

[189]  Guo, S.F., et al., Fe-based bulk metallic glasses: Brittle or ductile?. Applied Physics Letters, 2014. 105(16): 161901.

[190]  Gu, X.J., S.J. Poon and G.J. Shiflet, Mechanical properties of iron-based bulk metallic glasses. Journal of Materials Research, 2007. 22(02): 344–351.

[191]  Xi, X.K., et al., Fracture of brittle metallic glasses: brittleness or plasticity. Physical Review Letters, 2005. 94(12): 125510.

[192]  Chen, T.-H. and C.-K. Tsai, The microstructural evolution and mechanical properties of Zr-based metallic glass under different strain rate compressions. Materials, 2015. 8(4): 1831.

[193]  Xue, Y.F., et al., Deformation and failure behavior of a hydrostatically extruded Zr38Ti17Cu10.5Co12Be22.5 bulk metallic glass/porous tungsten phase composite under dynamic compression. Composites Science and Technology, 2008. 68(15–16): 3396–3400.

[194]  Chen, H.S., Plastic flow in metallic glasses under compression. Scripta Metallurgica, 1973. 7(9): 931–935.

[195]  Flores, K.M. and R.H. Dauskardt, Fracture and deformation of bulk metallic glasses and their composites. Intermetallics, 2004. 12(7–9): 1025–1029.

[196]    Lowhaphandu, P. and J.J. Lewandowski, Fracture toughness and notched toughness of bulk amorphous alloy: Zr-Ti-Ni-Cu-Be. Scripta Materialia, 1998. 38(12): 1811–1817.

[197]    Lowhaphandu, P., et al., Deformation and fracture toughness of a bulk amorphous Zr–Ti–Ni–Cu–Be alloy. Intermetallics, 2000. 8(5–6): 487–492.

[198]    Conner, R.D., et al., Fracture toughness determination for a beryllium-bearing bulk metallic glass. Scripta Materialia, 1997. 37(9): 1373–1378.

[199]    Gilbert, C.J., R.O. Ritchie and W.L. Johnson, Fracture toughness and fatigue-crack propagation in a Zr–Ti–Ni–Cu–Be bulk metallic glass. Applied Physics Letters, 1997. 71(4): 476–478.

[200]    Xu, J., U. Ramamurty and E. Ma, The fracture toughness of bulk metallic glasses. JOM, 2010. 62(4): 10–18.

[201]    Kimura, H. and T. Masumoto, Fracture toughness of amorphous metals. Scripta Metallurgica, 1975. 9(3): 211–221.

[202]    Chen, M., Mechanical Behavior of Metallic Glasses: Microscopic Understanding of Strength and Ductility. Annual Review of Materials Research, 2008. 38(1): 445–469.

[203]    Hufnagel, T.C., C.A. Schuh and M.L. Falk, Deformation of metallic glasses: Recent developments in theory, simulations, and experiments. Acta Materialia, 2016. 109: 375–393.

[204]    Sarac, B. and J. Schroers, Designing tensile ductility in metallic glasses. Nature Communications, 2013. 4: 2158.

[205]    Ritchie, R.O., The conflicts between strength and toughness. Nature Materials, 2011. 10(11): 817–822.

[206]    Wada, T., A. Inoue and A.L. Greer, Enhancement of room-temperature plasticity in a bulk metallic glass by finely dispersed porosity. Applied Physics Letters, 2005. 86(25): 251907.

[207]    Donovan, P.E. and W.M. Stobbs, Shear band interactions with crystals in partially crystallised metallic glasses. Journal of Non-Crystalline Solids, 1983. 55(1): 61–76.

[208]    Leng, Y. and T.H. Courtney, Multiple shear band formation in metallic glasses in composites. Journal of Materials Science, 1991. 26(3): 588–592.

[209]    Liu, L.F., et al., Behavior of multiple shear bands in Zr-based bulk metallic glass. Materials Chemistry and Physics, 2005. 93(1): 174–177.

[210]    Huang, R., et al., Inhomogeneous deformation in metallic glasses. Journal of the Mechanics and Physics of Solids, 2002. 50(5): 1011–1027.

[211]    Spaepen, F., A microscopic mechanism for steady state inhomogeneous flow in metallic glasses. Acta Metallurgica, 1977. 25(4): 407–415.

[212]    Steif, P.S., F. Spaepen and J.W. Hutchinson, Strain localization in amorphous metals. Acta Metallurgica, 1982. 30(2): 447–455.

[213]    Flores, K.M., Structural changes and stress state effects during inhomogeneous flow of metallic glasses. Scripta Materialia, 2006. 54(3): 327–332.

[214]    Donovan, P.E. and W.M. Stobbs, The structure of shear bands in metallic glasses. Acta Metallurgica, 1981. 29(8): 1419–1436.

[215]    Li, N., W. Chen and L. Liu, Thermoplastic micro-forming of bulk metallic glasses: a review. JOM, 2016. 68(4): 1246–1261.

[216]    Sarac, B., et al., Three-dimensional shell fabrication using blow molding of bulk metallic glass. Journal of Microelectromechanical Systems, 2011. 20(1): 28–36.

[217]    Schroers, J., The superplastic forming of bulk metallic glasses. JOM, 2005. 57(5): 35–39.

[218]    Dandliker, R.B., R.D. Conner and W.L. Johnson, Melt infiltration casting of bulk metallic-glass matrix composites. Journal of Materials Research, 1998. 13(10): 2896–2901.

[219]    Jiang, Y. and K. Qiu. Computational micromechanics analysis of toughening mechanisms of particle-reinforced bulk metallic glass composites. Materials & Design 1980–2015, 2015. 65: 410–416.

[220]   Sun, Y.F., et al., Formation, thermal stability and deformation behavior of graphite-flakes reinforced Cu-based bulk metallic glass matrix composites. Materials Science and Engineering: A, 2006. 435–436: 132–138.

[221]   Conner, R.D., R.B. Dandliker and W.L. Johnson, Mechanical properties of tungsten and steel fiber reinforced Zr41.25Ti13.75Cu12.5Ni10Be22.5 metallic glass matrix composites. Acta Materialia, 1998. 46(17): 6089–6102.

[222]   Lee, K., et al., Direct observation of microfracture process in metallic-continuous-fiber-reinforced amorphous matrix composites fabricated by liquid pressing process. Materials Science and Engineering: A, 2010. 527(4–5): 941–946.

[223]   Wadhwa, P., J. Heinrich and R. Busch, Processing of copper fiber-reinforced Zr41.2Ti13.8Cu12.5Ni10.0Be22.5 bulk metallic glass composites. Scripta Materialia, 2007. 56(1): 73–76.

[224]   Cytron, S.J., A metallic glass-metal matrix composite. Journal of Materials Science Letters, 1982. 1(5): 211–213.

[225]   Deng, S.T., et al., Metallic glass fiber-reinforced Zr-based bulk metallic glass. Scripta Materialia, 2011. 64(1): 85–88.

[226]   Wang, Z., et al., Microstructure and mechanical behavior of metallic glass fiber-reinforced Al alloy matrix composites. Scientific Reports, 2016. 6: 24384.

[227]   Jiang, J.-Z., et al., Low-density high-strength bulk metallic glasses and their composites: a review. Advanced Engineering Materials, 2015. 17(6): 761–780.

[228]   Qiao, J., In-situ dendrite/metallic glass matrix composites: a review. Journal of Materials Science & Technology, 2013. 29(8): 685–701.

[229]   Freed, R.L. and J.B. Vander Sande, The effects of devitrification on the mechanical properties of Cu46Zr54 metallic glass. Metallurgical Transactions A, 1979. 10(11): 1621–1626.

[230]   Greer, A.L., Y.Q. Cheng and E. Ma, Shear bands in metallic glasses. Materials Science and Engineering: R: Reports, 2013. 74(4): 71–132.

[231]   Gludovatz, B., et al., Size-dependent fracture toughness of bulk metallic glasses. Acta Materialia, 2014. 70: 198–207.

[232]   Ogata, S., et al., Atomistic simulation of shear localization in Cu–Zr bulk metallic glass. Intermetallics, 2006. 14(8–9): 1033–1037.

[233]   Packard, C.E. and C.A. Schuh, Initiation of shear bands near a stress concentration in metallic glass. Acta Materialia, 2007. 55(16): 5348–5358.

[234]   Pampillo, C.A., Localized shear deformation in a glassy metal. Scripta Metallurgica, 1972. 6(10): 915–917.

[235]   Zhou, M., A.J. Rosakis and G. Ravichandran, On the growth of shear bands and failure-mode transition in prenotched plates: A comparison of singly and doubly notched specimens. International Journal of Plasticity, 1998. 14(4): 435–451.

[236]   Pan, D., et al., Experimental characterization of shear transformation zones for plastic flow of bulk metallic glasses. Proceedings of the National Academy of Sciences, 2008. 105(39): 14769–14772.

[237]   Das, J., et al., "Work-hardenable" ductile bulk metallic glass. Physical Review Letters, 2005. 94(20): 205501.

[238]   Schroers, J. and W.L. Johnson, Ductile bulk metallic glass. Physical Review Letters, 2004. 93(25): 255506.

[239]   Jiang, W.H., et al., Ductility of a Zr-based bulk-metallic glass with different specimen's geometries. Materials Letters, 2006. 60(29–30): 3537–3540.

[240]   Das, J., et al., Plasticity in bulk metallic glasses investigated via the strain distribution. Physical Review B, 2007. 76(9): 092203.

[241]   Chen, L.Y., et al., New class of plastic bulk metallic glass. Physical Review Letters, 2008. 100(7): 075501.

[242]   Abdeljawad, F., M. Fontus and M. Haataja, Ductility of bulk metallic glass composites: Microstructural effects. Applied Physics Letters, 2011. 98(3): 031909.

[243]   Magagnosc, D.J., et al., Tunable Tensile Ductility in Metallic Glasses. Scientific Reports, 2013. 3: 1096.

[244]   Lu, X.L., et al., Gradient Confinement Induced Uniform Tensile Ductility in Metallic Glass. Scientific Reports, 2013. 3: 3319.

[245]   Yao, K.F., et al., Superductile bulk metallic glass. Applied Physics Letters, 2006. 88(12): 122106.

[246]   Song, K.K., et al., Strategy for pinpointing the formation of B2 CuZr in metastable CuZr-based shape memory alloys. Acta Materialia, 2011. 59(17): 6620–6630.

[247]   Ding, J., et al., Large-sized CuZr-based bulk metallic glass composite with enhanced mechanical properties. Journal of Materials Science & Technology, 2014. 30(6): 590–594.

[248]   Jiang, F., et al., Microstructure evolution and mechanical properties of Cu46Zr47Al7 bulk metallic glass composite containing CuZr crystallizing phases. Materials Science and Engineering: A, 2007. 467(1–2): 139–145.

[249]   Liu, J., et al., Microstructure and compressive properties of in-situ martensite CuZr phase reinforced ZrCuNiAl metallic glass matrix composite. Materials Transactions, 2010. 51(5): 1033–1037.

[250]   Liu, Z., et al., Microstructural tailoring and improvement of mechanical properties in CuZr-based bulk metallic glass composites. Acta Materialia, 2012. 60(6–7): 3128–3139.

[251]   Liu, Z.Q., et al., Microstructural percolation assisted breakthrough of trade-off between strength and ductility in CuZr-based metallic glass composites. Scientific Reports, 2014. 4: 4167.

[252]   Schryvers, D., et al., Unit cell determination in CuZr martensite by electron microscopy and X-ray diffraction. Scripta Materialia, 1997. 36(10): 1119–1125.

[253]   Seo, J.W. and D. Schryvers, TEM investigation of the microstructure and defects of CuZr martensite. Part I: Morphology and twin systems. Acta Materialia, 1998. 46(4): 1165–1175.

[254]   Seo, J.W. and D. Schryvers, TEM investigation of the microstructure and defects of CuZr martensite. Part II: Planar defects. Acta Materialia, 1998. 46(4): 1177–1183.

[255]   Song, K., Synthesis, microstructure, and deformation mechanisms of CuZr-based bulk metallic glass composites. 2013,

[256]   Song, K.K., et al., Correlation between the microstructures and the deformation mechanisms of CuZr-based bulk metallic glass composites. AIP Advances, 2013. 3(1): 012116.

[257]   Kim, D.H., et al., Phase separation in metallic glasses. Progress in Materials Science, 2013. 58(8): 1103–1172.

[258]   Sun, L., et al., Phase separation and microstructure evolution of Zr48Cu36Ag8Al8 bulk metallic glass in the supercooled liquid region. Rare Metal Materials and Engineering, 2016. 45(3): 567–570.

[259]   Antonowicz, J., et al., Early stages of phase separation and nanocrystallization in Al-rare earth metallic glasses studied using SAXS/WAXS and HRTEM methods. Reviews on Advanced Materials Science, 2008. 18(5): 454–458.

[260]   Park, B.J., et al., Phase separating bulk metallic glass: a hierarchical composite. Physical Review Letters, 2006. 96(24): 245503.

[261]   Kelton, K.F., A new model for nucleation in bulk metallic glasses. Philosophical Magazine Letters, 1998. 77(6): 337–344.

[262]   Chang, H.J., et al., Synthesis of metallic glass composites using phase separation phenomena. Acta Materialia, 2010. 58(7): 2483–2491.

[263]  Kündig, A.A., et al., In situ formed two-phase metallic glass with surface fractal microstructure. Acta Materialia, 2004. 52(8): 2441–2448.

[264]  Oh, J.C., et al., Phase separation in Cu43Zr43Al7Ag7 bulk metallic glass. Scripta Materialia, 2005. 53(2): 165–169.

[265]  Park, E.S. and D.H. Kim, Phase separation and enhancement of plasticity in Cu–Zr–Al–Y bulk metallic glasses. Acta Materialia, 2006. 54(10): 2597–2604.

[266]  Guo, G.-Q., et al., How can synchrotron radiation techniques be applied for detecting microstructures in amorphous alloys?. Metals, 2015. 5(4): 2048.

[267]  Guo, G.-Q., et al., Detecting structural features in metallic glass via synchrotron radiation experiments combined with simulations. Metals, 2015. 5(4): 2093.

[268]  Michalik, S., et al., Structural modifications of swift-ion-bombarded metallic glasses studied by high-energy X-ray synchrotron radiation. Acta Materialia, 2014. 80: 309–316.

[269]  Mu, J., et al., In situ high-energy X-ray diffraction studies of deformation-induced phase transformation in Ti-based amorphous alloy composites containing ductile dendrites. Acta Materialia, 2013. 61(13): 5008–5017.

[270]  Paradis, P.-F., et al., Materials properties measurements and particle beam interactions studies using electrostatic levitation. Materials Science and Engineering: R: Reports, 2014. 76: 1–53.

[271]  Huang, Y.J., J. Shen and J.F. Sun, Bulk metallic glasses: Smaller is softer. Applied Physics Letters, 2007. 90(8): 081919.

[272]  Oh, Y.S., et al., Microstructure and tensile properties of high-strength high-ductility Ti-based amorphous matrix composites containing ductile dendrites. Acta Materialia, 2011. 59(19): 7277–7286.

[273]  Kim, K.B., et al., Heterogeneous distribution of shear strains in deformed Ti66.1Cu8Ni4.8Sn7.2Nb13.9 nanostructure-dendrite composite. Physica Status Solidi (a), 2005. 202(13): 2405–2412.

[274]  He, G., et al., Novel Ti-base nanostructure-dendrite composite with enhanced plasticity. Nature Materials, 2003. 2(1): 33–37.

[275]  Wang, Y., et al., Investigation of the microcrack evolution in a Ti-based bulk metallic glass matrix composite. Progress in Natural Science: Materials International, 2014. 24(2): 121–127.

[276]  Zhang, T., et al., Dendrite size dependence of tensile plasticity of in situ Ti-based metallic glass matrix composites. Journal of Alloys and Compounds, 2014. 583: 593–597.

[277]  Gargarella, P., et al., Ti–Cu–Ni shape memory bulk metallic glass composites. Acta Materialia, 2013. 61(1): 151–162.

[278]  Hofmann, D.C., et al., New processing possibilities for highly toughened metallic glass matrix composites with tensile ductility. Scripta Materialia, 2008. 59(7): 684–687.

[279]  Chu, J.P., Annealing-induced amorphization in a glass-forming thin film. JOM, 2009. 61(1): 72–75.

[280]  Kingery, W.D., Introduction to Ceramics. 1960, Wiley.

[281]  Conner, R.D., H. Choi-Yim and W.L. Johnson, Mechanical properties of Zr57Nb5Al10Cu15.4Ni12.6 metallic glass matrix particulate composites. Journal of Materials Research, 1999. 14(08): 3292–3297.

[282]  Choi-Yim, H., et al., Quasistatic and dynamic deformation of tungsten reinforced Zr57Nb5Al10Cu15.4Ni12.6 bulk metallic glass matrix composites. Scripta Materialia, 2001. 45(9): 1039–1045.

[283]  Hui, X., et al., Wetting angle and infiltration velocity of Zr base bulk metallic glass composite. Intermetallics, 2006. 14(8–9): 931–935.

[284]  Xue, Y.F., et al., Strength-improved Zr-based metallic glass/porous tungsten phase composite by hydrostatic extrusion. Applied Physics Letters, 2007. 90(8): 081901.

[285]   Hou, B., et al., Dynamic and quasi-static mechanical properties of fibre-reinforced metallic glass at different temperatures. Philosophical Magazine Letters, 2007. 87(8): 595–601.

[286]   Zhang, Y. and A.L. Greer, Correlations for predicting plasticity or brittleness of metallic glasses. Journal of Alloys and Compounds, 2007. 434–435: 2–5.

[287]   Fu, X.L., Y. Li and C.A. Schuh, Temperature, strain rate and reinforcement volume fraction dependence of plastic deformation in metallic glass matrix composites. Acta Materialia, 2007. 55(9): 3059–3071.

[288]   Argon, A.S., Plastic deformation in metallic glasses. Acta Metallurgica, 1979. 27(1): 47–58.

[289]   Chen, Y., et al., Preparation, microstructure and deformation behavior of Zr-based metallic glass/porous SiC interpenetrating phase composites. Materials Science and Engineering: A, 2011. 530: 15–20.

[290]   Chen, C., et al., Effect of temperature on the dynamic mechanical behaviors of Zr-based metallic glass reinforced porous tungsten matrix composite. Advanced Engineering Materials, 2012. 14(7): 439–444.

[291]   Liu, T., et al., Microstructures and Mechanical Properties of ZrC Reinforced (Zr-Ti)-Al-Ni-Cu Glassy Composites by an In Situ Reaction. Advanced Engineering Materials, 2009. 11(5): 392–398.

[292]   Zhang, H.F., et al., Synthesis and characteristics of 80 vol.% tungsten (W) fibre/Zr based metallic glass composite. Intermetallics, 2009. 17(12): 1070–1077.

[293]   Khademian, N. and R. Gholamipour, Fabrication and mechanical properties of a tungsten wire reinforced Cu–Zr–Al bulk metallic glass composite. Materials Science and Engineering: A, 2010. 527(13–14): 3079–3084.

[294]   Ott, R.T., et al., Micromechanics of deformation of metallic-glass–matrix composites from in situ synchrotron strain measurements and finite element modeling. Acta Materialia, 2005. 53(7): 1883–1893.

[295]   Stoica, M., et al., Mechanical response of metallic glasses: Insights from in-situ high energy X-ray diffraction. JOM, 2010. 62(2): 76–82.

[296]   Choi-Yim, H. and W.L. Johnson, Bulk metallic glass matrix composites. Applied Physics Letters, 1997. 71(26): 3808–3810.

[297]   Hays, C.C., C.P. Kim and W.L. Johnson, Improved mechanical behavior of bulk metallic glasses containing in situ formed ductile phase dendrite dispersions. Materials Science and Engineering: A, 2001. 304–306: 650–655.

[298]   Zhuang, S., J. Lu and G. Ravichandran, Shock wave response of a zirconium-based bulk metallic glass and its composite. Applied Physics Letters, 2002. 80(24): 4522–4524.

[299]   Löser, W., et al., Effect of casting conditions on dendrite-amorphous/nanocrystalline Zr–Nb–Cu–Ni–Al in situ composites. Intermetallics, 2004. 12(10–11): 1153–1158.

[300]   Liu, Z., et al., Pronounced ductility in CuZrAl ternary bulk metallic glass composites with optimized microstructure through melt adjustment. AIP Advances, 2012. 2(3): 032176.

[301]   Eckert, J., et al., Structural bulk metallic glasses with different length-scale of constituent phases. Intermetallics, 2002. 10(11–12): 1183–1190.

[302]   Das, J., et al., Designing bulk metallic glass and glass matrix composites in martensitic alloys. Journal of Alloys and Compounds, 2009. 483(1–2): 97–101.

[303]   Wu, D., et al., Deformation-induced martensitic transformation in Cu-Zr-Zn bulk metallic glass composites. Metals, 2015. 5(4): 2134.

[304]   Pauly, S., et al., Deformation-induced martensitic transformation in Cu–Zr–(Al,Ti) bulk metallic glass composites. Scripta Materialia, 2009. 60(6): 431–434.

[305]   Javid, F.A., et al., Martensitic transformation and thermal cycling effect in Cu–Co–Zr alloys. Journal of Alloys and Compounds, 2011. 509(Supplement 1:): S334–S337.

[306]   Prashanth, K.G., et al., Production of high strength Al85Nd8Ni5Co2 alloy by selective laser melting. Additive Manufacturing, 2015. 6: 1–5.

[307]   Jung, H.Y., et al., Fabrication of Fe-based bulk metallic glass by selective laser melting: A parameter study. Materials & Design, 2015. 86: 703–708.

[308]   Park, J.M., et al., High-strength bulk Al-based bimodal ultrafine eutectic composite with enhanced plasticity. Journal of Materials Research, 2009. 24(08): 2605–2609.

[309]   He, G., J. Eckert and W. Löser, Stability, phase transformation and deformation behavior of Ti-base metallic glass and composites. Acta Materialia, 2003. 51(6): 1621–1631.

[310]   Das, J., et al., High-strength Ti-base ultrafine eutectic with enhanced ductility. Applied Physics Letters, 2005. 87(16): 161907.

[311]   Wu, F.F., et al., Strength asymmetry of ductile dendrites reinforced Zr- and Ti-based composites. Journal of Materials Research, 2006. 21(09): 2331–2336.

[312]   Eckert, J., et al., Mechanical properties of bulk metallic glasses and composites. Journal of Materials Research, 2007. 22(02): 285–301.

[313]   Wang, G., et al., Plastic flow of a Cu50Zr45Ti5 bulk metallic glass composite. Journal of Materials Science & Technology, 2014. 30(6): 609–615.

[314]   Sun, B.A., et al., Serrated flow and stick–slip deformation dynamics in the presence of shear-band interactions for a Zr-based metallic glass. Acta Materialia, 2012. 60(10): 4160–4171.

[315]   Sun, B.A., et al., Origin of intermittent plastic flow and instability of shear band sliding in bulk metallic glasses. Physical Review Letters, 2013. 110(22): 225501.

[316]   Qu, R.T., et al., Metallic glasses: Notch-insensitive materials. Scripta Materialia, 2012. 66(10): 733–736.

[317]   Pauly, S., et al., Modeling deformation behavior of Cu–Zr–Al bulk metallic glass matrix composites. Applied Physics Letters, 2009. 95(10): 101906.

[318]   Antonaglia, J., et al., Bulk metallic glasses deform via slip avalanches. Physical Review Letters, 2014. 112(15): 155501.

[319]   Hufnagel, T.C., et al., Controlling shear band behavior in metallic glasses through microstructural design. Intermetallics, 2002. 10(11–12): 1163–1166.

[320]   Hufnagel, T.C., Preface to the viewpoint set on mechanical behavior of metallic glasses. Scripta Materialia, 2006. 54(3): 317–319.

[321]   Hufnagel, T.C., U.K. Vempati and J.D. Almer, Crack-tip strain field mapping and the toughness of metallic glasses. PLoS ONE, 2013. 8(12): e83289.

[322]   Georgarakis, K., et al., Shear band melting and serrated flow in metallic glasses. Applied Physics Letters, 2008. 93(3): 031907.

[323]   Qiao, J.W., Y. Zhang and P.K. Liaw, Serrated flow kinetics in a Zr-based bulk metallic glass. Intermetallics, 2010. 18(11): 2057–2064.

[324]   Cunliffe, A., et al., Glass formation in a high entropy alloy system by design. Intermetallics, 2012. 23: 204–207.

[325]   Hufnagel, T.C., R.T. Ott and J. Almer, Structural aspects of elastic deformation of a metallic glass. Physical Review B, 2006. 73(6): 064204.

[326]   Hufnagel, T.C., et al., Deformation and failure of Zr57Ti5Cu20Ni8Al10 bulk metallic glass under quasi-static and dynamic compression. Journal of Materials Research, 2002. 17(06): 1441–1445.

[327]   Ott, R.T., et al., Yield criteria and strain-rate behavior of Zr57.4Cu16.4Ni8.2Ta8Al10 metallic-glass-matrix composites. Metallurgical and Materials Transactions A, 2006. 37(11): 3251–3258.

[328]   An, Z.N., et al., Interface constraints on shear band patterns in bonded metallic glass films under microindentation. Metallurgical and Materials Transactions A, 2012. 43(8): 2729–2741.

[329]  Antonaglia, J., et al., Tuned critical avalanche scaling in bulk metallic glasses. Scientific Reports, 2014. 4: 4382.

[330]  Chu, J.P., et al., Thin film metallic glasses: Unique properties and potential applications. Thin Solid Films, 2012. 520(16): 5097–5122.

[331]  Chu, J.P., et al., Thin film metallic glasses: Preparations, properties, and applications. JOM, 2010. 62(4): 19–24.

[332]  Du, X.H., et al., Two-glassy-phase bulk metallic glass with remarkable plasticity. Applied Physics Letters, 2007. 91(13): 131901.

[333]  Fan, C., et al., Mechanical behavior of bulk amorphous alloys reinforced by ductile particles at cryogenic temperatures. Physical Review Letters, 2006. 96(14): 145506.

[334]  Jiang, W.H., et al., Spatiotemporally inhomogeneous plastic flow of a bulk-metallic glass. International Journal of Plasticity, 2008. 24(1): 1–16.

[335]  Miller, M.K. and P. Liaw, Bulk Metallic Glasses: An Overview. 2007, Springer US.

[336]  Qiao, J.W., et al., Characteristic of improved fatigue performance for Zr-based bulk metallic glass matrix composites. Materials Science and Engineering: A, 2013. 563: 101–105.

[337]  Qiao, J.W., et al., Distinguished work-hardening capacity of a Ti-based metallic glass matrix composite upon dynamic loading. Materials Science and Engineering: A, 2013. 585: 277–280.

[338]  Qiao, J.W., et al., Dynamic shear punching of metallic glass matrix composites. Intermetallics, 2013. 36: 31–35.

[339]  Qiao, J.W., et al., Micromechanisms of plastic deformation of a dendrite/Zr-based bulk-metallic-glass composite. Scripta Materialia, 2009. 61(11): 1087–1090.

[340]  Wang, G.Y., P.K. Liaw and M.L. Morrison, Progress in studying the fatigue behavior of Zr-based bulk-metallic glasses and their composites. Intermetallics, 2009. 17(8): 579–590.

[341]  Wang, Y.-C., et al., Indentation behavior of Zr-based metallic-glass films via molecular-dynamics simulations. Metallurgical and Materials Transactions A, 2010. 41(11): 3010–3017.

[342]  Huang, J.C., J.P. Chu and J.S.C. Jang, Recent progress in metallic glasses in Taiwan. Intermetallics, 2009. 17(12): 973–987.

[343]  Jung, H.Y., et al., Crystallization kinetics of Fe76.5–x C6.0Si3.3B5.5P8.7Cu x (x = 0, 0.5, and 1 at. pct) bulk amorphous alloy. Metallurgical and Materials Transactions A, 2015. 46(6): 2415–2421.

[344]  Rim, K.R., et al., Tensile necking and enhanced plasticity of cold rolled β-Ti dendrite reinforced Ti-based bulk metallic glass matrix composite. Journal of Alloys and Compounds, 2013. 579: 253–258.

[345]  Jeon, C., et al., Effects of dendrite size on tensile deformation behavior in Zr-based amorphous matrix composites containing ductile dendrites. Metallurgical and Materials Transactions A, 2012. 43(10): 3663–3674.

[346]  Porter, D.A. and K.E. Easterling, Phase Transformations in Metals and Alloys, Third Edition (Revised Reprint). 1992, Taylor & Francis.

[347]  Park, E.S., J.S. Kyeong and D.H. Kim, Phase separation and improved plasticity by modulated heterogeneity in Cu–(Zr, Hf)–(Gd, Y)–Al metallic glasses. Scripta Materialia, 2007. 57(1): 49–52.

[348]  Van De Moortèle, B., et al., Phase separation before crystallization in Zr-Ti-Cu-Ni-Be bulk metallic glasses: Influence of the chemical composition. Journal of Non-Crystalline Solids, 2004. 345–346: 169–172.

[349]  Bracchi, A., et al., Decomposition and metastable phase formation in the bulk metallic glass matrix composite Zr56Ti14Nb5Cu7Ni6Be12. Journal of Applied Physics, 2006. 99(12): 123519.

[350]  Inoue, A., et al., Production methods and properties of engineering glassy alloys and composites. Intermetallics, 2015. 58: 20–30.

[351]  Wang, B., et al., Simulation of solidification microstructure in twin-roll casting strip. Computational Materials Science, 2010. 49(1, Supplement): S135–S139.

[352]  Inoue, A. and T. Zhang, Fabrication of bulky Zr-based glassy alloys by suction casting into copper mold. Materials Transactions, JIM, 1995. 36(9): 1184–1187.

[353]  Inoue, A. and T. Zhang, Fabrication of bulk glassy $Zr_{55}Al_{10}Ni_5Cu_{30}$ alloy of 30 mm in diameter by a suction casting method. Materials Transactions, JIM, 1996. 37(2): 185–187.

[354]  Lou, H.B., et al., 73 mm-diameter bulk metallic glass rod by copper mould casting. Applied Physics Letters, 2011. 99(5): 051910.

[355]  Qiao, J.W., et al., Synthesis of plastic Zr-based bulk metallic glass matrix composites by the copper-mould suction casting and the Bridgman solidification. Journal of Alloys and Compounds, 2009. 477(1–2): 436–439.

[356]  Wall, J.J., et al., A combined drop/suction-casting machine for the manufacture of bulk-metallic-glass materials. Review of Scientific Instruments, 2006. 77(3): 033902.

[357]  Figueroa, I., et al., Preparation of Cu-based bulk metallic glasses by suction casting. 2007.

[358]  Zhu, Z., et al., Fabricating Zr-based bulk metallic glass microcomponent by suction casting using silicon micromold. Advances in Mechanical Engineering, 2014. 6: 362484.

[359]  Hofmann, D.C., et al., Semi-solid induction forging of metallic glass matrix composites. JOM, 2009. 61(12): 11–17.

[360]  Khalifa, H.E., Bulk metallic glasses and their composites: composition optimization, Thermal Stability, and Microstructural Tunability. 2009,

[361]  Hofmann, D.C. and S.N. Roberts, Microgravity metal processing: from undercooled liquids to bulk metallic glasses. Npj Microgravity, 2015. 1: 15003.

[362]  Wang, D., et al., Superplastic micro-forming of Zr65Cu17.5Ni10Al7.5 bulk metallic glass with silicon mold using hot embossing technology. Journal of Alloys and Compounds, 2009. 484 (1–2): 118–122.

[363]  Jeong, H.G., S.J. Yoo and W.J. Kim, Micro-forming of Zr65Al10Ni10Cu15 metallic glasses under superplastic condition. Journal of Alloys and Compounds, 2009. 483(1–2): 283–285.

[364]  Duan, G., et al., Bulk metallic glass with benchmark thermoplastic processability. Advanced Materials, 2007. 19(23): 4272–4275.

[365]  Ramasamy, P., et al., High pressure die casting of Fe-based metallic glass. Scientific Reports, 2016. 6(1): 35258.

[366]  Liu, L., et al., Near-net forming complex shaped Zr-based bulk metallic glasses by high pressure die casting. Materials, 2018. 11(11.

[367]  Liu, L.H., et al., Determination of forming ability of high pressure die casting for Zr-based metallic glass. Journal of Materials Processing Technology, 2017. 244: 87–96.

[368]  Inoue, A., et al., Bulky amorphous alloys produced by a high-pressure die casting process. Key Engineering Materials, 1993. 81–83: 147–152.

[369]  Inoue, A., et al., Mg-Cu-Y bulk amorphous alloys with high tensile strength produced by a high-pressure die casting method. 1992.

[370]  Inoue, A., et al., Bulky La–Al–TM (TM=transition metal) amorphous alloys with high tensile strength produced by a high-pressure die casting method. Materials Transactions, JIM, 1993. 34(4): 351–358.

[371]  Bai, L., et al., Experimental and numerical simulations of the solidification process in continuous casting of slab. Metals, 2016. 6(3): 53.

[372]  Haag, F., et al., Assessing continuous casting of precious bulk metallic glasses. Journal of Non-Crystalline Solids, 2019. 521: 119120.

[373]  Pei, Z. and D. Ju, Simulation of the continuous casting and cooling behavior of metallic glasses. Materials, 2017. 10(4).

[374]   Kang, H.G., et al., Fabrication of bulk Mg–Cu–Ag–Y glassy alloy by squeeze casting. Materials Transactions, JIM, 2000. 41(7): 846–849.

[375]   Nishida, Y. and H. Matsubara, Effect of Pressure on Heat Transfer at the Metal Mould-Casting Interface. Br Foundryman, 1976. 69: 274–278.

[376]   Wiest, A., Injection molding metallic glass. Scripta Mater., 2009. 60.

[377]   Zhang, N., et al., Replication of micro/nano-scale features by micro injection molding with a bulk metallic glass mold insert. Journal of Micromechanics and Microengineering, 2012. 22(6): 065019.

[378]   He, P., et al., Bulk metallic glass mold for high volume fabrication of micro optics. Microsystem Technologies, 2016. 22(3): 617–623.

[379]   Zhang, N., et al., Towards nano-injection molding. Materials Today, 2012. 15(5): 216–221.

[380]   R, N. and B. R, Fabrication of bulk metallic glasses by centrifugal casting method. Journal of Achievements in Materials and Manufacturing Engineering, 2007. 20.

[381]   Kumar, R., R. Kumar, S. Chattopadhyaya, A. Ghosh and A. Kumar (2015) Friction stir welding of BMG's: a review.

[382]   Jamili-Shirvan, Z., et al., Mechanical and thermal properties of identified zones at a Ti-based bulk metallic glass weld spot jointed by friction stir spot welding (FSSW). Journal of Non-Crystalline Solids, 2020. 544: 120188.

[383]   Ma, X., S.M. Howard and B.K. Jasthi, Friction stir welding of bulk metallic glass Vitreloy 106a. Journal of Manufacturing Science and Engineering, 2014. 136(5.

[384]   Zhang, H., et al., Joining of Zr 51 Ti 5 Ni 10 Cu 25 Al 9 BMG to aluminum alloy by friction stir welding. Vacuum, 2015. 120.

[385]   Ji, Y.S., et al., Friction stir welding of Zr-based bulk metallic glass. Journal of Physics: Conference Series, 2009. 165: p. 012015.

[386]   Kobata, J., et al., Nanoscale amorphous "band-like" structure induced by friction stir processing in Zr55Cu30Al10Ni5 bulk metallic glass. Materials Letters, 2007. 61(17): 3771–3773.

[387]   Li, F.P., et al., Microstructure and mechanical properties of friction stir welded joint of Zr46Cu46Al8 bulk metallic glass with pure aluminum. Materials Science and Engineering: A, 2013. 588: 196–200.

[388]   Zhang, H., et al., Joining of Zr51Ti5Ni10Cu25Al9 BMG to aluminum alloy by friction stir welding. Vacuum, 2015. 120: 47–49.

[389]   Sun, Y., et al., Microstructure and mechanical properties of friction stir welded joint of Zr55Cu30Al10Ni5 bulk metallic glass with pure copper. Materials Science and Engineering: A, 2010. 527(15): 3427–3432.

[390]   Wang, D., et al., Friction stir welding of Zr55Cu30Al10Ni5 bulk metallic glass to Al–Zn–Mg–Cu alloy. Scripta Materialia, 2009. 60(2): 112–115.

[391]   Chiba, A., Y. Kawamura and M. Nishida, Explosive welding of ZrTiCuNiBe bulk metallic glass to crystalline metallic plates. Materials Science Forum, 2008. 566: 119–124.

[392]   Kawamura, Y., Y. Ohno and A. Chiba, Development of welding technologies in bulk metallic glasses. Materials Science Forum, 2002. 386–388: 553–558.

[393]   Feng, J., P. Chen and Q. Zhou, Investigation on explosive welding of Zr53Cu35Al12 bulk metallic glass with crystalline copper. Journal of Materials Engineering and Performance, 2018. 27(6): 2932–2937.

[394]   Jiang, M.Q., et al., Joining of bulk metallic glass to brass by thick-walled cylinder explosion. Scripta Materialia, 2015. 97: 17–20.

[395]   Liang, H., et al., Experimental and numerical simulation study of Zr-based BMG/Al composites manufactured by underwater explosive welding. Journal of Materials Research and Technology, 2020. 9(2): 1539–1548.

[396] Ji, H., L. Li and M. Li, Low-temperature joining of Fe-based amorphous foil with aluminum by ultrasonic-assisted soldering with Sn-based fillers. Materials & Design, 2015. 84: 254–260.

[397] Tamura, S., et al., Ultrasonic cavitation treatment for soldering on Zr-based bulk metallic glass. Journal of Materials Processing Technology, 2008. 206.

[398] Jing, Y., et al., The influence of Zr content on the performance of TiZrCuNi brazing filler. Materials Science and Engineering: A, 2016. 678: 190–196.

[399] Lee, M.K. and J.G. Lee, Mechanical and corrosion properties of Ti–6Al–4V alloy joints brazed with a low-melting-point 62.7Zr–11.0Ti–13.2Cu–9.8Ni–3.3Be amorphous filler metal. Materials Characterization, 2013. 81: 19–27.

[400] Liu, S., et al., Interfacial microstructure and shear strength of TC4 alloy joints vacuum brazed with Ti–Zr–Ni–Cu filler metal. Materials Science and Engineering: A, 2020. 775: 138990.

[401] Liu, Y.H., et al., Microstructural and mechanical properties of jointed ZrO2/Ti–6Al–4V alloy using Ti33Zr17Cu50 amorphous brazing filler. Materials & Design, 2013. 47: 281–286.

[402] Tariq, N.H., et al., Electron beam brazing of Zr62Al13Ni7Cu18 bulk metallic glass with Ti metal. Vacuum, 2014. 101.

[403] Dong, H., et al., CuTiNiZrV Amorphous Alloy Foils for Vacuum Brazing of TiAl Alloy to 40Cr Steel. Journal of Materials Science & Technology, 2015. 31(2): 217–222.

[404] Kim, J. and T. Lee, Brazing method to join a novel Cu54Ni6Zr22Ti18 bulk metallic glass to carbon steel. Sci Technol Weld Join, 2017. 22.

[405] Sypien, A., et al., Effect of Pd, temperature and time on wetting and interfacial microstructure of bulk metallic glasses TiCuZrPd on Ti-6Al-4V substrate. Journal of Alloys and Compounds, 2017. 695: 962–970.

[406] Pilarczyk, W., Structure and properties of Zr-based bulk metallic glasses in As-cast state and after laser welding. Materials (Basel), 2018. 11(7). https://doi.org/10.3390/ma11071117.

[407] Caiazzo, F. and V. Alfieri, Optimization of Laser Beam Welding Of Steel Parts made by Additive Manufacturing. 2021, The International Journal of Advanced Manufacturing Technology.

[408] Chen, B., et al., Crystallization of Zr55Cu30Al10Ni5 bulk metallic glass in laser welding: simulation and experiment. Advanced Engineering Materials, 2015. 17.

[409] Chen, B., et al., Laser welding of annealed Zr55Cu30Ni5Al10 bulk metallic glass. Intermetallics, 2014. 46.

[410] Pilarczyk, W., A Study of Structure and Morphology of Bulk Metallic Glasses after Laser Beam Treatment. Journal of Materials Engineering and Performance, 2020. 29(3): 1460–1466.

[411] Wang, G., et al., Laser welding of Ti40Zr25Ni3Cu12Be20 bulk metallic glass. Materials Science and Engineering, 2012. 541.

[412] Wang, H.-S., H.-G. Chen and J.S.-C. Jang, Microstructure evolution in Nd: YAGlaser-welded (Zr53Cu30Ni9Al8)Si0.5 bulk metallic glass alloy. Journal of Alloys and Compounds, 2010. 495(1): 224–228.

[413] Wang, H.S., et al., Combination of a Nd: YAGlaser and a liquid cooling device to (Zr53Cu30Ni9Al8)Si0.5 bulk metallic glass welding. Materials Science and Engineering, 2010. 528.

[414] Xing, W., et al., Effect of energy density on defect evolution in 3D printed Zr-based metallic glasses by selective laser melting. Science China Physics, Mechanics & Astronomy, 2019. 63(2): 226111.

[415] Zhang, W., et al., Effect of laser surface melting on bulk metallic glass: Investigation of microstructure, microhardness, friction and wear properties. Journal of Alloys and Compounds, 2018. 732: 792–798.

[416]   Kawamura, Y. and Y. Ohno, Successful Electron-Beam Welding of Bulk Metallic Glass. Materials Transactions, 2001. 42(11): 2476–2478.

[417]   Kim, J., Weldability of Cu54Zr22Ti18Ni6 bulk metallic glass by ultrasonic welding processing. Materials Letters, 2014. 130: 160–163.

[418]   Kreye, H., M. Hammerschmidt and G. Reiners, Ultrasonic welding of metallic alloy glass to copper. Scripta Metallurgica, 1978. 12(11): 1059–1061.

[419]   Kumar, S., et al., Application of ultrasonic vibrations in welding and metal processing: A status review. Journal of Manufacturing Processes, 2017. 26: 295–322.

[420]   Li, X., et al., Cold joining to fabricate large size metallic glasses by the ultrasonic vibrations. Scripta Materialia, 2020. 185: 100–104.

[421]   Song, X.C., Z.Q. Zhu and Y.F. Chen, Ultrasonic welding of Fe78Si9B13 metallic glass. Materials Science Forum, 2015. 809–810: 348–353.

[422]   Shin, H., J. Park and Y. Yokoyama, Dissimilar friction welding of tubular Zr-based bulk metallic glasses. The Journal of Alloys and Compounds, 2010. 504.

[423]   Shin, H.S., et al., Joining of Zr-based bulk metallic glasses using the friction welding method. The Journal of Alloys and Compounds, 2007. 434–435.

[424]   Shoji, T., Y. Kawamura and Y. Ohno, Friction welding of bulk metallic glasses to different ones. Materials Science and Engineering A, 2004. 375–377.

[425]   Li, F.P., et al., Microstructure and mechanical properties of friction stir welded joint of Zr46Cu46Al8 bulk metallic glass with pure aluminum. Materials Science and Engineering A, 2013. 588.

[426]   Mahmoudiniya, M. and L.A.I. Kestens, Microstructural development and texture evolution in the stir zone and thermomechanically affected zone of a ferrite-martensite steel friction stir weld. Materials Characterization, 2021. 175: 111053.

[427]   Phillips, B.J., et al., Effect of parallel deposition path and interface material flow on resulting microstructure and tensile behavior of Al-Mg-Si alloy fabricated by additive friction stir deposition. Journal of Materials Processing Technology, 2021. 295: 117169.

[428]   Shin, H., Tool geometry effect on the characteristics of dissimilar friction stir spot welded bulk metallic glass to lightweight alloys. Journal of Alloys and Compounds , 2014. 586.

[429]   Chiba, A., et al., Explosive welding of ZrTiCuNiBe bulk metallic glass to crystalline Cu plate. Materials Science Forum, 2011. 673.

[430]   Liu, W.D., et al., Metallic glass coating on metals plate by adjusted explosive welding technique. Applied Surface Science, 2009. 255(23): 9343–9347.

[431]   Fukumoto, S., et al., Estimation of current path area during small scale resistance spot welding of bulk metallic glass to stainless steel. Science and Technology of Welding and Joining, 2013. 18(2): 135–142.

[432]   Gao, X.F., et al., Deformation and fracture of a Zr-Al-Cu metallic glass ribbon under tension near glass transition temperature. China Foundry, 2018. 15.

[433]   Makhanlall, D., et al., Joining of Ti-based bulk metallic glasses using resistance spot welding technology. The Journal of Materials Processing Technology, 2012. 212.

[434]   Kim, J., S. Shin and C. Lee, Characterization of the Gas Tungsten Arc Welded $Cu_{54}Ni_6Zr_{22}Ti_{18}$ Bulk Metallic Glass Weld. Materials Transactions, 2005. 46(6): 1440–1442.

[435]   Schroers, J., C. Veazey and W.L. Johnson, Amorphous metallic foam. Applied Physics Letters, 2003. 82(3): 370–372.

[436]   Brothers, A.H. and D.C. Dunand, Syntactic bulk metallic glass foam. Applied Physics Letters, 2004. 84.

[437]   Kai, W., Oxidation behavior of a Pd43Cu27Ni10P20 bulk metallic glass and foam in dry air. Metallurgical and Materials Transactions A, 2010. 41.

[438] Li, J.B., et al., Novel open-cell bulk metallic glass foams with promising characteristics. Materials Letters, 2013. 105.

[439] Nguyen, V.T., et al., Synthesis of biocompatible TiZr-based bulk metallic glass foams for bio-implant application. Materials Letters, 2019. 256: 126650.

[440] Lin, H., et al., A metallic glass syntactic foam with enhanced energy absorption performance. Scripta Materialia, 2016. 119: 47–50.

[441] Lee, M.H. and D.J. Sordelet, Synthesis of bulk metallic glass foam by powder extrusion with a fugitive second phase. Applied Physics Letters, 2006. 89(2): 021921.

[442] Apreutesei, M., et al., Zr–Cu thin film metallic glasses: An assessment of the thermal stability and phases' transformation mechanisms. Journal of Alloys and Compounds, 2015. 619: 284–292.

[443] Chu, J.P., Thin film metallic glasses: Unique properties and potential applications. Thin Solid Films, 2012. 520.

[444] Diyatmika, W., et al., Thin film metallic glass as an underlayer for tin whisker mitigation: A room-temperature evaluation. Thin Solid Films, 2014. 561: 93–97.

[445] Lou, B.-S., et al., Corrosion property and biocompatibility evaluation of Fe–Zr–Nb thin film metallic glasses. Thin Solid Films, 2019. 691: 137615.

[446] Mohri, M., et al., Thermal stability of the Ti-Zr-Cu-Pd nano-glassy thin films. Journal of Alloys and Compounds, 2018. 735: 2197–2204.

[447] Chu, C.W., et al., Study of the characteristics and corrosion behavior for the Zr-based metallic glass thin film fabricated by pulse magnetron sputtering process. Thin Solid Films, 2009. 517.

[448] Das, S., et al., Electromechanical behavior of pulsed laser deposited platinum-based metallic glass thin films. Physica Status Solidi (a), 2016. 213(2): 399–404.

[449] Glasscott, M.W., et al., Electrosynthesis of high-entropy metallic glass nanoparticles for designer, multi-functional electrocatalysis. Nature Communications, 2019. 10.

[450] Gu, J., et al., Novel corrosion behaviours of the annealing and cryogenic thermal cycling treated Ti-based metallic glasses. Intermetallics, 2019. 110: 106467.

[451] Lee, J., M.-L. Liou and J.-G. Duh, The development of a Zr-Cu-Al-Ag-N thin film metallic glass coating in pursuit of improved mechanical, corrosion, and antimicrobial property for bio-medical application. Surface and Coatings Technology, 2017. 310: 214–222.

[452] Sarac, B., et al., Electrocatalytic behavior of hydrogenated Pd-metallic glass nanofilms: Butler-Volmer, Tafel, and Impedance Analyses. Electrocatalysis, 2020. 11(1): 94–109.

[453] Wu, Z.F., et al., Thickness-dependent pitting corrosion behavior in Ni–Nb thin film metallic glass. Thin Solid Films, 2014. 564: 294–298.

[454] Zhou, P., et al., Electrodeposition of FeCoP nanoglass films. Microelectronic Engineering, 2020. 229: 111363.

[455] Chang, C.H., et al., Fatigue property improvements of ZK60 magnesium alloy: Effects of thin film metallic glass. Thin Solid Films, 2016. 616: 431–436.

[456] Lee, C.M., et al., Fatigue property improvements of Ti–6Al–4V by thin film coatings of metallic glass and TiN: a comparison study. Thin Solid Films, 2014. 561: 33–37.

[457] Tsai, P.H., et al., Fatigue properties improvement of high-strength aluminum alloy by using a ZrCu-based metallic glass thin film coating. Thin Solid Films, 2014. 561: 28–32.

[458] Tsai, P.H., et al., Coating thickness effect of metallic glass thin film on the fatigue-properties improvement of 7075 aluminum alloy. Thin Solid Films, 2019. 677: 68–72.

[459] Yi, S.M., et al., Electrical reliability and interfacial adhesion of Cu(Mg) thin films for interconnect process adaptability. Thin Solid Films, 2008. 516.

[460]  Wu, C.-Y., Y.-C. Wang and C. Chen, Indentation properties of Cu–Zr–Al metallic-glass thin films at elevated temperatures via molecular dynamics simulation. Computational Materials Science, 2015. 102: 234–242.

[461]  Barbee, T.W., et al., Synthesis of amorphous niobium-nickel alloys by vapor quenching. Thin Solid Films, 1977. 45.

[462]  Chen, H. and C. Miller, A rapid quenching technique for the preparation of thin uniform films of amorphous solids. Review of Scientific Instruments, 1970. 41.

[463]  Chuang, C.-Y., et al., Mechanical properties study of a magnetron-sputtered Zr-based thin film metallic glass. Surface and Coatings Technology, 2013. 215: 312–321.

[464]  Tillmann, W., et al., LPBF-M manufactured Zr-based bulk metallic glasses coated with magnetron sputtered ZrN films. Surface and Coatings Technology, 2020. 386: 125463.

[465]  Thanka Rajan, S., et al., Materials properties of ion beam sputtered Ti-Cu-Pd-Zr thin film metallic glasses. Journal of Non-Crystalline Solids, 2017. 461: 104–112.

[466]  Liu, M.C., et al., Superplastic-like deformation in metallic amorphous/crystalline nanolayered micropillars. Intermetallics, 2012. 30: 30–34.

[467]  Li, X.P., et al., Selective laser melting of an Al86Ni6Y4.5Co2La1.5 metallic glass: Processing, microstructure evolution and mechanical properties. Materials Science and Engineering: A, 2014. 606: 370–379.

[468]  Li, X.P., et al., The role of a low-energy–density re-scan in fabricating crack-free Al85Ni5Y6Co2Fe2 bulk metallic glass composites via selective laser melting. Materials & Design, 2014. 63: 407–411.

[469]  Li, X.P., et al., Effect of substrate temperature on the interface bond between support and substrate during selective laser melting of Al–Ni–Y–Co–La metallic glass, Materials & Design 1980–2015, 2015. 65: 1–6.

[470]  Kruth, J.P., et al., Selective laser melting of iron-based powder. Journal of Materials Processing Technology, 2004. 149(1–3): 616–622.

[471]  Thijs, L., et al., A study of the microstructural evolution during selective laser melting of Ti–6Al–4V. Acta Materialia, 2010. 58(9): 3303–3312.

[472]  Vastola, G., et al., Modeling the microstructure evolution during additive manufacturing of Ti6Al4V: A comparison between electron beam melting and selective laser melting. JOM, 2016. 68(5): 1370–1375.

[473]  Zhang, L.C., et al., Manufacture by selective laser melting and mechanical behavior of a biomedical Ti–24Nb–4Zr–8Sn alloy. Scripta Materialia, 2011. 65(1): 21–24.

[474]  Dutta, B. and F.H. Froes, Chapter 1 – The Additive Manufacturing of Titanium Alloys, in Additive Manufacturing of Titanium Alloys.. 2016, Butterworth-Heinemann. 1–10.

[475]  Chen, B., et al., Improvement in mechanical properties of a Zr-based bulk metallic glass by laser surface treatment. Journal of Alloys and Compounds, 2010. 504,(Supplement 1:): S45–S47.

[476]  Santos, E.C., et al., Rapid manufacturing of metal components by laser forming. International Journal of Machine Tools and Manufacture, 2006. 46(12–13): 1459–1468.

[477]  Sun, H. and K.M. Flores, Laser deposition of a Cu-based metallic glass powder on a Zr-based glass substrate. Journal of Materials Research, 2008. 23(10): 2692–2703.

[478]  Sun, H. and K. Flores, Microstructural analysis of a laser-processed Zr-based bulk metallic glass. Metallurgical and Materials Transactions A, 2010. 41(7): 1752–1757.

[479]  Sun, H. and K.M. Flores, Spherulitic crystallization mechanism of a Zr-based bulk metallic glass during laser processing. Intermetallics, 2013. 43: 53–59.

[480]  Welk, B.A., et al., Phase Selection in a Laser Surface Melted Zr-Cu-Ni-Al-Nb Alloy. Metallurgical and Materials Transactions B, 2014. 45(2): 547–554.

[481]   Borkar, T., et al., A combinatorial assessment of AlxCrCuFeNi2 (0 < x < 1.5) complex
        concentrated alloys: Microstructure, microhardness. and magnetic properties. Acta
        Materialia, 2016. 116: 63–76.

[482]   Thompson, S.M., et al., An overview of direct laser deposition for additive manufacturing;
        Part I: Transport phenomena, modeling and diagnostics. Additive Manufacturing, 2015.
        8: 36–62.

[483]   Shamsaei, N., et al., An overview of direct laser deposition for additive manufacturing; Part
        II: Mechanical behavior, process parameter optimization and control. Additive
        Manufacturing, 2015. 8: 12–35.

[484]   Ding, D., et al., Wire-feed additive manufacturing of metal components: technologies,
        developments and future interests. The International Journal of Advanced Manufacturing
        Technology, 2015. 81(1): 465–481.

[485]   Welk, B.A., M.A. Gibson and H.L. Fraser, A combinatorial approach to the investigation of
        metal systems that form both bulk metallic glasses and high entropy alloys. JOM, 2016.
        68(3): 1021–1026.

[486]   Hofmann, D.C., et al., Compositionally graded metals: A new frontier of additive
        manufacturing. Journal of Materials Research, 2014. 29(17): 1899–1910.

[487]   Hofmann, D.C., et al., Developing gradient metal alloys through radial deposition additive
        manufacturing. Scientific Reports, 2014. 4: 5357.

[488]   Joseph, J., et al., Comparative study of the microstructures and mechanical properties of
        direct laser fabricated and arc-melted AlxCoCrFeNi high entropy alloys. Materials Science
        and Engineering: A, 2015. 633: 184–193.

[489]   Ocelík, V., et al., Additive manufacturing of high-entropy alloys by laser processing. JOM,
        2016. 68(7): 1810–1818.

[490]   Tsai, P. and K.M. Flores, A laser deposition strategy for the efficient identification of glass-
        forming alloys. Metallurgical and Materials Transactions A, 2015. 46(9): 3876–3882.

[491]   Sun, Y.F., et al., Effect of Nb content on the microstructure and mechanical properties of
        Zr–Cu–Ni–Al–Nb glass forming alloys. Journal of Alloys and Compounds, 2005. 403(1–2):
        239–244.

[492]   Kühn, U., et al., Microstructure and mechanical properties of slowly cooled Zr–Nb–Cu–Ni–
        Al composites with ductile bcc phase. Materials Science and Engineering: A, 2004.
        375–377: 322–326.

[493]   Sun, Y.F., et al., Brittleness of Zr-based bulk metallic glass matrix composites containing
        ductile dendritic phase. Materials Science and Engineering: A, 2005. 406(1–2): 57–62.

[494]   Sun, Y.F., et al., Effect of quasicrystalline phase on the deformation behavior of
        Zr62Al9.5Ni9.5Cu14Nb5 bulk metallic glass. Materials Science and Engineering: A, 2005.
        398(1–2): 22–27.

[495]   Kühn, U., et al., ZrNbCuNiAl bulk metallic glass matrix composites containing dendritic bcc
        phase precipitates. Applied Physics Letters, 2002. 80(14): 2478–2480.

[496]   Firstov, G.S., J. Van Humbeeck and Y.N. Koval, High-temperature shape memory alloys:
        Some recent developments. Materials Science and Engineering: A, 2004. 378(1–2): 2–10.

[497]   Nishida, M., et al., New deformation twinning mode of B19′ martensite in Ti-Ni shape
        memory alloy. Scripta Materialia, 1998. 39(12): 1749–1754.

[498]   Schryvers, D., et al., Applications of advanced transmission electron microscopic
        techniques to Ni–Ti based shape memory materials. Materials Science and Engineering:
        A, 2004. 378(1–2): 11–15.

[499]   Lee, J.-C., et al., Strain hardening of an amorphous matrix composite due to deformation-
        induced nanocrystallization during quasistatic compression. Applied Physics Letters,
        2004. 84(15): 2781–2783.

[500]  Shi, Y. and M.L. Falk, Stress-induced structural transformation and shear banding during simulated nanoindentation of a metallic glass. Acta Materialia, 2007. 55(13): 4317–4324.

[501]  Louzguine-Luzgin, D.V., et al., High-strength and ductile glassy-crystal Ni–Cu–Zr–Ti composite exhibiting stress-induced martensitic transformation. Philosophical Magazine, 2009. 89(32): 2887–2901.

[502]  Hao, S., et al., A Transforming Metal Nanocomposite with Large Elastic Strain, Low Modulus, and High Strength. Science, 2013. 339(6124): 1191–1194.

[503]  Yang, Y. and C.T. Liu, Size effect on stability of shear-band propagation in bulk metallic glasses: an overview. Journal of Materials Science, 2012. 47(1): 55–67.

[504]  Jiang, W.H. and M. Atzmon, Mechanically-assisted nanocrystallization and defects in amorphous alloys: A high-resolution transmission electron microscopy study. Scripta Materialia, 2006. 54(3): 333–336.

[505]  Dodd, B. and Y. Bai, Adiabatic Shear Localization: Frontiers and Advances. 2012, Elsevier.

[506]  Jiang, W.H., F.E. Pinkerton and M. Atzmon, Deformation-induced nanocrystallization: A comparison of two amorphous Al-based alloys. Journal of Materials Research, 2005. 20(03): 696–702.

[507]  Kelton, K. and A.L. Greer, Nucleation in Condensed Matter: Applications in Materials and Biology. 2010, Elsevier Science.

[508]  Germain, P. and K. Zellama, Nucleation in Condensed Matter, in Cohesive Properties of Semiconductors under Laser Irradiation, Laude, L.D.,Editor. 1983, Springer Netherlands: Dordrecht. 254–278.

[509]  Fecht, H.J. and W.L. Johnson, Thermodynamic properties and metastability of bulk metallic glasses. Materials Science and Engineering: A, 2004. 375–377: 2–8.

[510]  Chen, Q., et al., A new criterion for evaluating the glass-forming ability of bulk metallic glasses. Materials Science and Engineering: A, 2006. 433(1): 155–160.

[511]  Rashidi, R., M. Malekan and R. Gholamipour, Crystallization kinetics of Cu47Zr47Al6 and (Cu47Zr47Al6)99Sn1 bulk metallic glasses. Journal of Non-Crystalline Solids, 2018. 498: 272–280.

[512]  Shao, L., et al., Effects of Si addition on glass-forming ability and crystallization behavior of DyCoAl bulk metallic glass. Journal of Alloys and Compounds, 2021. 874: 159964.

[513]  Sohn, S.W., et al., Phase separation in bulk-type Gd–Zr–Al–Ni metallic glass. Intermetallics, 2012. 23: 57–62.

[514]  Yuan, -Z.-Z., et al., A new criterion for evaluating the glass-forming ability of bulk glass forming alloys. Journal of Alloys and Compounds, 2008. 459(1): 251–260.

[515]  Lu, S., et al., The effect of Y addition on the crystallization behaviors of Zr-Cu-Ni-Al bulk metallic glasses. Journal of Alloys and Compounds, 2019. 799: 501–512.

[516]  Mohammadi Rahvard, M., M. Tamizifar and S.M.A. Boutorabi, The effect of Ag addition on the non-isothermal crystallization kinetics and fragility of Zr56Co28Al16 bulk metallic glass. Journal of Non-Crystalline Solids, 2018. 481: 74–84.

[517]  Sohrabi, S. and R. Gholamipour, Effect of Nb minor addition on the crystallization kinetics of Zr-Cu-Al-Ni metallic glass. Journal of Non-Crystalline Solids, 2021. 560: 120731.

[518]  Bai, F.X., et al., Crystallization kinetics of an Au-based metallic glass upon ultrafast heating and cooling. Scripta Materialia, 2017. 132: 58–62.

[519]  Biroli, G., et al., Thermodynamic signature of growing amorphous order in glass-forming liquids. Nat. Phys, 2008. 4.

[520]  Browne, D.J., Z. Kovacs and W.U. Mirihanage, Comparison of nucleation and growth mechanisms in alloy solidification to those in metallic glass crystallisation – relevance to modeling. Transactions of the Indian Institute of Metals, 2009. 62(4): 409–412.

[521]  Wang, Z., et al., Nucleation and thermal stability of an icosahedral nanophase during the early crystallization stage in Zr-Co-Cu-Al metallic glasses. Acta Materialia, 2017. 132: 298–306.

[522]  Louzguine-Luzgin, D.V., et al., Glass-transition process in an Au-based metallic glass. Journal of Non-Crystalline Solids, 2015. 419: 12–15.

[523]  Zhang, J.Y., G. Liu and J. Sun, Self-toughening crystalline Cu/amorphous Cu–Zr nanolaminates: Deformation-induced devitrification. Acta Materialia, 2014. 66: 22–31.

[524]  El-Eskandarany, M.S., J. Saida and A. Inoue, Room-temperature mechanically induced solid-state devitrifications of glassy Zr65Al7.5Ni10Cu12.5Pd5 alloy powders. Acta Materialia, 2003. 51(15): 4519–4532.

[525]  Liu, H., et al., Crystallization in additive manufacturing of metallic glasses: A review. Additive Manufacturing, 2020. 36: 101568.

[526]  Wang, Y.Q., et al., Size- and constituent-dependent deformation mechanisms and strain rate sensitivity in nanolaminated crystalline Cu/amorphous Cu–Zr films. Acta Materialia, 2015. 95: 132–144.

[527]  Fernandes, D.J., et al., Mechanical strength and surface roughness of magnesium-based metallic glasses. JOM, 2017. 69(7): 1175–1184.

[528]  Abrosimova, G., et al., Devitrification of Zr55Cu30Al15Ni5Bulk metallic glass under heating and HPT deformation. Metals, 2020. 10(10).

[529]  Fujita, K., et al., Fatigue properties in high strength bulk metallic glasses. Intermetallics, 2012. 30: 12–18.

[530]  Jia, H., et al., Fatigue and fracture behavior of bulk metallic glasses and their composites. Progress in Materials Science, 2018. 98: 168–248.

[531]  Luo, J., et al., Low-cycle fatigue of metallic glass nanowires. Acta Materialia, 2015. 87: 225–232.

[532]  Wang, X.D., et al., Improving fatigue property of metallic glass by tailoring the microstructure to suppress shear band formation. Materialia, 2019. 7: 100407.

[533]  Kumar, A., et al., Optimization of mechanical and corrosion properties of plasma sprayed low-chromium containing Fe-based amorphous/nanocrystalline composite coating. Surface and Coatings Technology, 2019. 370: 255–268.

[534]  Rafique, M., et al., Effect of single pass laser surface treatment on microstructure evolution of inoculated Zr 47.5 Cu 45.5 Al 5 Co 2 and non-inoculated Zr 65 Cu 15 Al 10 Ni 10 bulk metallic glass matrix composites. Engineering, 2018. 10: 730–758.

[535]  Rafique, M.M.A., Additive manufacturing of bulk metallic glasses and their composites– recent trends and approaches.

[536]  Rafique, M.M.A., et al., Effect of laser additive manufacturing on microstructure evolution of inoculated Zr47. 5Cu45. 5Al5Co2 bulk metallic glass matrix composites.

[537]  Dey, G.K. and S. Banerjee, Role of diffusion in glass formation and crystallization in metallic glasses. 1999.

[538]  Faupel, F., Diffusion in metallic glasses and supercooled melts. Reviews of Modern Physics, 2003. 75.

[539]  Faupel, F., et al., Diffusion in metallic glasses and supercooled melts. Reviews of Modern Physics, 2003. 75(1): 237–280.

[540]  Pogatscher, S., et al., Monotropic polymorphism in a glass-forming metallic alloy. Journal of Physics: Condensed Matter, 2018. 30(23): 234002.

[541]  Yin, Z., et al., Polyamorphism in a solute-lean Al–Ce metallic glass. Journal of Applied Physics, 2021. 129(2): 025108.

[542] Tellería, I. and J.M. Barandiarán, Kinetics of the primary, eutectic and polymorphic crystallization of metallic glasses studied by continuous scan methods. Thermochimica Acta, 1996. 280–281: 279–287.

[543] Ketov, S.V., et al., High-resolution transmission electron microscopy investigation of diffusion in metallic glass multilayer films. Materials Today Advances, 2019. 1: 100004.

[544] Li, Z., et al., Forming of metallic glasses: mechanisms and processes. Materials Today Advances, 2020. 7: 100077.

[545] Johnson, W.L., J.H. Na and M.D. Demetriou, Quantifying the origin of metallic glass formation. Nature Communications, 2016. 7(1): 10313.

[546] Hu, Z., et al., Oxidation feature and diffusion mechanism of Zr-based metallic glasses near the glass transition point. Materials Research Express, 2018. 5(3): 036511.

[547] Schawe, J.E.K., S. Pogatscher and J.F. Löffler, Thermodynamics of polymorphism in a bulk metallic glass: Heat capacity measurements by fast differential scanning calorimetry. Thermochimica Acta, 2020. 685: 178518.

[548] Perim, E., et al., Predicting bulk metallic glass forming ability with the thermodynamic density of competing crystalline states. arXiv preprint arXiv:1601.08233, 2016.

[549] Perim, E., et al., Spectral descriptors for bulk metallic glasses based on the thermodynamics of competing crystalline phases. Nature Communications, 2016. 7(1): 12315.

[550] Ruta, B., E. Pineda and Z. Evenson, Relaxation processes and physical aging in metallic glasses. Journal of Physics: Condensed Matter, 2017. 29(50): 503002.

[551] Liu, B.B., et al., Viscosity, relaxation and fragility of the Ca65Mg15Zn20 bulk metallic glass. Intermetallics, 2019. 109: 8–15.

[552] Atzmon, M. and J.D. Ju, Microscopic description of flow defects and relaxation in metallic glasses. Physical Review E, 2014. 90(4): 042313.

[553] Haruyama, O., et al., Volume and enthalpy relaxation in Zr55Cu30Ni5Al10 bulk metallic glass. Acta Materialia, 2010. 58(5): 1829–1836.

[554] Haruyama, O., et al., Comparison of structural relaxation behavior in as-cast and pre-annealed Zr-based bulk metallic glasses just below glass transition. Materials Transactions, 2015. 56(5): 648–654.

[555] Guo, W., R. Yamada and J. Saida, Unusual plasticization for structural relaxed bulk metallic glass. Materials Science and Engineering: A, 2017. 699: 81–87.

[556] Greer, A.L. and F. Spaepen, Creep, diffusion, and structural relaxation in metallic glasses*. Annals of the New York Academy of Sciences, 1981. 371(1): 218–237.

[557] Wang, Q., et al., Unusual fast secondary relaxation in metallic glass. Nature Communications, 2015. 6(1): 7876.

[558] Bohmer, R., Correlation of primary and secondary relaxations in a supercooled liquid. Physical Review Letters, 2006. 97.

[559] He, N., et al., The evolution of relaxation modes during isothermal annealing and its influence on properties of Fe-based metallic glass. Journal of Non-Crystalline Solids, 2019. 509: 95–98.

[560] Hu, L. and Y.Z. Yue, Secondary relaxation in metallic glass formers: its correlation with the genuine Johari-Goldstein relaxation. Journal of Physical Chemistry, 2009. 113.

[561] Pelletier, J.M., et al., Main and secondary relaxations in an Au-based bulk metallic glass investigated by mechanical spectroscopy. Journal of Alloys and Compounds, 2016. 684: 530–536.

[562] Stevenson, J.D. and P.G. Wolynes, A universal origin for secondary relaxations in supercooled liquids and structural glasses. Nature Physics, 2009. 6.

[563]    Xiao, C.D., J.Y. Jho and A.F. Yee, Correlation between the shear yielding behavior and secondary relaxations of bisphenol a polycarbonate and related copolymers. Macromolecules, 1994. 27.

[564]    Shimono, M. and H. Onodera, Molecular dynamics study on structural relaxation of metallic glasses. Materials Science Forum, 2010. 638–642: 1648–1652.

[565]    Il'enko, V.V. and V.V. Sviridov, Kinetics of structural relaxation in glasses: A hierarchical model of relaxation centers. Glass Physics and Chemistry, 2000. 26(3): 252–256.

[566]    Kosilov, A.T. and V.A. Khonik, Directed structural relaxation and homogeneous flow of fresh quenching metal glasses. Izvestiya Akademii Nauk – Rossijskaya Akademiya Nauk Seriya Fizicheskaya, 1993. 57(11): 192–197.

[567]    Khonik, V.A., Understanding of the structural relaxation of metallic glasses within the framework of the interstitialcy theory. Metals, 2015. 5(2).

[568]    Khonik, V.A., et al., The role of structural relaxation in the plastic flow of metallic glasses. Journal of Applied Physics, 1998. 83(11): 5724–5731.

[569]    Louzguine-Luzgin, D.V., et al., Influence of cyclic loading on the structure and double-stage structure relaxation behavior of a Zr-Cu-Fe-Al metallic glass. Materials Science and Engineering: A, 2019. 742: 526–531.

[570]    Afonin, G.V., et al., Structural relaxation and related viscous flow of Zr–Cu–Al-based bulk glasses produced from the melts with different glass-forming ability. Intermetallics, 2011. 19(9): 1298–1305.

[571]    Castellero, A., B. Moser and D.I. Uhlenhaut, Room-temperature creep and structural relaxation of Mg-Cu-Y metallic glasses. Acta Mater, 2008. 56.

[572]    Evenson, Z., et al., The effect of low-temperature structural relaxation on free volume and chemical short-range ordering in a Au49Cu26.9Si16.3Ag5.5Pd2.3 bulk metallic glass. Scripta Materialia, 2015. 103: 14–17.

[573]    Granato, A.V. and V.A. Khonik, An interstitialcy theory of structural relaxation and related viscous flow of glasses. Physical Review Letters, 2004. 93.

[574]    Hwang, J., et al., Nanoscale structure and structural relaxation in Zr 50 Cu 45 Al 5 bulk metallic glass. Physical Review Letters, 2012. 108(19): 195505.

[575]    Schoenholz, S.S., et al., A structural approach to relaxation in glassy liquids. Nature Physics, 2016. 12.

[576]    Slipenyuk, A. and J. Eckert, Correlation between enthalpy change and free volume reduction during structural relaxation of Zr55Cu30Al10Ni5 metallic glass. Scripta Materialia, 2004. 50.

[577]    Slipenyuk, A. and J. Eckert, Correlation between enthalpy change and free volume reduction during structural relaxation of Zr55Cu30-Al10Ni5 metallic glass. Scripta Mater, 2004. 50.

[578]    Wang, T., Y.Q. Yang and J.B. Li, Thermodynamics and structural relaxation in Ce-based bulk metallic glass-forming liquids. Journal of Alloys and Compounds, 2011. 509.

[579]    Wang, W.H., Dynamic relaxations and relaxation-property relationships in metallic glasses. Progress in Materials Science, 2019. 106: 100561.

[580]    Ouyang, S., et al., Chemical independent relaxation in metallic glasses from the nanoindentation experiments. Journal of Applied Physics, 2017. 121(24): 245104.

[581]    Zhang, J., et al., Composition dependent relaxation in La-Al-Ni/Cu metallic glasses. Journal of Alloys and Compounds, 2017. 726: 1024–1029.

[582]    Liu, C., E. Pineda and D. Crespo, Mechanical relaxation of metallic glasses: an overview of experimental data and theoretical models. Metals, 2015. 5(2).

[583]    Cheng, Y.T., et al., Effect of minor addition on dynamic mechanical relaxation in ZrCu-based metallic glasses. Journal of Non-Crystalline Solids, 2021. 553: 120496.

[584]  Hao, Q., W.Y. Jia and J.C. Qiao, Dynamic mechanical relaxation in Zr65Ni7Cu18Al10 metallic glass. Journal of Non-Crystalline Solids, 2020. 546: 120266.

[585]  Qiao, J.C., et al., Effects of iron addition on the dynamic mechanical relaxation of Zr55Cu30Ni5Al10 bulk metallic glasses. Journal of Alloys and Compounds, 2018. 749: 262–267.

[586]  Yao, Z.F., et al., Characterization and modeling of dynamic relaxation of a Zr-based bulk metallic glass. Journal of Alloys and Compounds, 2017. 690: 212–220.

[587]  Zhang, L.T., et al., Dynamic mechanical relaxation behavior of Zr35Hf17.5Ti5.5Al12.5Co7.5Ni12Cu10 high entropy bulk metallic glass. Journal of Materials Science & Technology, 2021. 83: 248–255.

[588]  Research Square, 2021.

[589]  Qiao, J.C., et al., Main α relaxation and slow β relaxation processes in a La30Ce30Al15Co25 metallic glass. Journal of Materials Science & Technology, 2019. 35(6): 982–986.

[590]  Qiao, J.C., R. Casalini and J.M. Pelletier, Main (α) relaxation and excess wing in Zr50Cu40Al10 bulk metallic glass investigated by mechanical spectroscopy. Journal of Non-Crystalline Solids, 2015. 407: 106–109.

[591]  Qiao, J.C. and J.M. Pelletier, Dynamic universal characteristic of the main (α) relaxation in bulk metallic glasses. Journal of Alloys and Compounds, 2014. 589: 263–270.

[592]  Pineda, E., et al., Relaxation of rapidly quenched metallic glasses: Effect of the relaxation state on the slow low temperature dynamics. Acta Materialia, 2013. 61(8): 3002–3011.

[593]  Zhai, W., et al., Distinctive slow β relaxation and structural heterogeneity in (LaCe)-based metallic glass. Journal of Alloys and Compounds, 2018. 742: 536–541.

[594]  Yu, H.-B., W.-H. Wang and K. Samwer, The β relaxation in metallic glasses: an overview. Materials Today, 2013. 16(5): 183–191.

[595]  Wang, D.P., J.C. Qiao and C.T. Liu, Relating structural heterogeneity to β relaxation processes in metallic glasses. Materials Research Letters, 2019. 7(8): 305–311.

[596]  Yu, H.B., et al., The β-relaxation in metallic glasses. National Science Review, 2014. 1(3): 429–461.

[597]  Cicerone, M.T. and J.F. Douglas, β-Relaxation governs protein stability in sugar-glass matrices. Soft Matter, 2012. 8.

[598]  Cui, X., et al., Influence of the chemical composition on the β-relaxation and the mechanical behavior of LaCe-based bulk metallic glasses. Journal of Non-Crystalline Solids, 2021. 562: 120779.

[599]  Cui, X., et al., Room temperature activated slow β relaxation and large compressive plasticity in a LaCe-based bulk metallic glass. Intermetallics, 2020. 122: 106793.

[600]  Jiang, W., J. Wu and B. Zhang, Size effect on beta relaxation in a La-based bulk metallic glass. Physica B: Condensed Matter, 2017. 509: 46–49.

[601]  Lee, J.-C., Calorimetric study of β-relaxation in an amorphous alloy: An experimental technique for measuring the activation energy for shear transformation. Intermetallics, 2014. 44: 116–120.

[602]  Ngai, K.L., Nature and properties of the Johari–Goldstein β-relaxation in the equilibrium liquid state of a class of glass-formers. The Journal of Chemical Physics, 2001. 115.

[603]  Wang, Z., et al., Pronounced slow β-relaxation in La-based bulk metallic glasses. Journal of Physics: Condensed Matter, 2011. 23.

[604]  Xu, H.Y. and M.Z. Li, Activation-relaxation technique study on β-relaxation in La55Ni20Al25 and Cu46Zr46Al8 metallic glasses. Intermetallics, 2018. 94: 10–16.

[605]  Yu, H.B., Tensile plasticity in metallic glasses with pronounced β relaxations. Physical Review Letters, 2012. 108.

[606]  Yu, H.B., et al., Chemical influence on β-relaxations and the formation of molecule-like metallic glasses. Nature Communications , 2013. 4.

[607]  Yu, H.B., et al., Correlation between β relaxation and self-diffusion of the smallest constituting atoms in metallic glasses. Physical Review Letters, 2012. 109.

[608]  Yu, H.B., et al., Relating activation of shear transformation zones to β relaxations in metallic glasses. Physical Review B, 2010. 81.

[609]  Yu, H.B., et al., The β -relaxation in metallic glasses. National Science Review, 2014. 1.

[610]  Yu, H.B., W.H. Wang and K. Samwer, The β relaxation in metallic glasses: an overview. Materials Today, 2013. 16.

[611]  Yu, H.B., et al., Relation between β relaxation and fragility in LaCe-based metallic glasses. Journal of Non-Crystalline Solids, 2012. 358.

[612]  Zhang, Y.R., et al., Relating activation of brittle-to-ductile transition to β relaxation in Cu46Zr44Al7Y3 metallic glass. Journal of Non-Crystalline Solids, 2020. 544: 120189.

[613]  Yu, H.B., et al., Tensile plasticity in metallic glasses with pronounced relaxations. Physical Review Letters, 2012. 108(1): 015504.

[614]  Zhu, Z.G., Compositional origin of unusual beta-relaxation properties in La-Ni-Al metallic glasses. The Journal of Chemical Physics, 2014. 141.

[615]  Das, A., et al., Stress breaks universal aging behavior in a metallic glass. Nature Communications, 2019. 10(1): 5006.

[616]  Küchemann, S. and R. Maaß, Gamma relaxation in bulk metallic glasses. Scripta Materialia, 2017. 137: 5–8.

[617]  Qiao, J., et al., Characteristics of the structural and Johari–Goldstein relaxations in Pd-based metallic glass-forming liquids. The Journal of Physical Chemistry B, 2014. 118(13): 3720–3730.

[618]  Qiao, J., et al., Characteristics of the structural and Johari–Goldstein relaxations in Pd-based metallic glass-forming liquids. The Journal of Physical Chemistry B, 2014. 118.

[619]  Qiao, J., J.-M. Pelletier and R. Casalini, Relaxation of bulk metallic glasses studied by mechanical spectroscopy. The Journal of Physical Chemistry B, 2013. 117(43): 13658–13666.

[620]  Zhong, C., et al., Deformation behavior of metallic glasses with shear band like atomic structure: a molecular dynamics study. Scientific Reports, 2016. 6(1): 30935.

[621]  Zhang, W., et al., Relaxation-to-rejuvenation transition of a Ce-based metallic glass by quenching/cryogenic treatment performed at sub-Tg. Journal of Alloys and Compounds, 2020. 825: 153997.

[622]  Cahn, R.W., et al., Studies of relaxation of metallic glasses by dilatometry and density measurements. MRS Proceedings, 1983. 28: 241.

[623]  Qiao, J.C., et al., Characteristics of stress relaxation kinetics of La 60 Ni 15 Al 25 bulk metallic glass. Acta Mater, 2015. 98.

[624]  Utiarahman, A., et al., Role of elastostatic loading and cyclic cryogenic treatment on relaxation behavior of Ce-based amorphous alloy. Materials Today Communications, 2021. 26: 101843.

[625]  Apreutesei, M., A. Billard and P. Steyer, Crystallization and hardening of Zr-40at.% Cu thin film metallic glass: Effects of isothermal annealing. Materials & Design, 2015. 86: 555–563.

[626]  Song, L., et al., Two-step relaxations in metallic glasses during isothermal annealing. Intermetallics, 2018. 93: 101–105.

[627]  Avchaciov, K.A., et al., Controlled softening of Cu64Zr36 metallic glass by ion irradiation. Applied Physics Letters, 2013. 102(18): 181910.

[628]  Magagnosc, D.J., et al., Effect of ion irradiation on tensile ductility, strength and fictive temperature in metallic glass nanowires. Acta Materialia, 2014. 74: 165–182.

[629]    Mayr, S.G., Impact of ion irradiation on the thermal, structural, and mechanical properties of metallic glasses. Physical Review B, 2005. 71(14): 144109.

[630]    Miglierini, M. and M. Hasiak, Impact of ion irradiation upon structure and magnetic properties of NANOPERM-type amorphous and nanocrystalline alloys. Journal of Nanomaterials, 2015. 2015: 175407.

[631]    Shao, L., et al., Mechanism of nanocrystallization temperature shifting during ion irradiation of metallic glasses. Nuclear Instruments and Methods in Physics Research Section B: Beam Interactions with Materials and Atoms, 2021. 497: 28–33.

[632]    Sun, K., et al., Structural rejuvenation and relaxation of a metallic glass induced by ion irradiation. Scripta Materialia, 2020. 180: 34–39.

[633]    Zhu, B., et al., Enhanced ductility in Cu64Zr36 metallic glasses induced by prolonged low-energy ion irradiation: A molecular dynamics study. Journal of Alloys and Compounds, 2021. 873: 159785.

[634]    Lv, Z., et al., Defects activation in CoFe-based metallic glasses during creep deformation. Journal of Materials Science & Technology, 2021. 69: 42–47.

[635]    Wang, T., et al., Structural heterogeneity originated plasticity in Zr–Cu–Al bulk metallic glasses. Intermetallics, 2020. 121: 106790.

[636]    Huang, Y., et al., The effects of annealing on the microstructure and the dynamic mechanical strength of a ZrCuNiAl bulk metallic glass. Intermetallics, 2013. 42: 192–197.

[637]    Qiao, J.C., et al., Understanding of micro-alloying on plasticity in Cu46Zr47–xAl7Dyx (0≤ x ≤ 8) bulk metallic glasses under compression: Based on mechanical relaxations and theoretical analysis. International Journal of Plasticity, 2016. 82: 62–75.

[638]    Wang, B., et al., Revealing the ultra-low-temperature relaxation peak in a model metallic glass. Acta Materialia, 2020. 195: 611–620.

[639]    Pan, J., et al., Extreme rejuvenation and softening in a bulk metallic glass. Nature Communications, 2018. 9(1): 560.

[640]    Ebner, C., et al., Effect of mechanically induced structural rejuvenation on the deformation behaviour of CuZr based bulk metallic glass. Materials Science and Engineering: A, 2020. 773: 138848.

[641]    Meng, F., et al., Reduction of shear localization through structural rejuvenation in Zr–Cu–Al bulk metallic glass. Materials Science and Engineering: A, 2019. 765: 138304.

[642]    Meng, F.Q., et al., Reversible transition of deformation mode by structural rejuvenation and relaxation in bulk metallic glass. Applied Physics Letters, 2012. 101.

[643]    Qiang, J. and K. Tsuchiya, Composition dependence of mechanically-induced structural rejuvenation in Zr-Cu-Al-Ni metallic glasses. Journal of Alloys and Compounds, 2017. 712: 250–255.

[644]    Qiang, J. and K. Tsuchiya, Concurrent solid-state amorphization and structural rejuvenation in Zr-Cu-Al alloy by high-pressure torsion. Materials Letters, 2017. 204: 138–140.

[645]    Tong, Y., et al., Structural rejuvenation in bulk metallic glasses. Acta Materialia, 2015. 86: 240–246.

[646]    Wang, D.P., Y. Yang and X.R. Niu, Resonance ultrasonic actuation and local structural rejuvenation in metallic glasses. Phys Rev B, 2017. 95.

[647]    Tong, Y., et al., Mechanical rejuvenation in bulk metallic glass induced by thermo-mechanical creep. Acta Materialia, 2018. 148: 384–390.

[648]    Guo, W., et al., Thermal rejuvenation of an Mg-based metallic glass. Metallurgical and Materials Transactions A, 2019. 50.

[649]    Guo, W., et al., Thermal rejuvenation of a heterogeneous metallic glass. Journal of Non-Crystalline Solids, 2018. 498: 8–13.

[650]  Guo, W., et al., Wu SS. Thermal rejuvenation of a heterogeneous metallic glass. Journal of Non-Crystalline Solids, 2018. 498.

[651]  Kim, Y.H., et al., Quenching temperature and cooling rate effects on thermal rejuvenation of metallic glasses. 2020, Metals and Materials International.

[652]  Song, W., et al., Improving plasticity of the Zr46Cu46Al8 bulk metallic glass via thermal rejuvenation. Science Bulletin, 2018. 63(13): 840–844.

[653]  Wakeda, M. and J. Saida, Heterogeneous structural changes correlated to local atomic order in thermal rejuvenation process of Cu-Zr metallic glass. Science and Technology of Advanced Materials, 2019. 20(1): 632–642.

[654]  Wang, C., et al., Rejuvenation of metallic glasses under high pressure. 2016.

[655]  Meng, Y., et al., Rejuvenation by enthalpy relaxation in metallic glasses. 2020.

[656]  Ketov, S.V., et al., Rejuvenation of metallic glasses by non-affine thermal strain. Nature, 2015. 524(7564): 200–203.

[657]  Ding, G., et al., Ultrafast extreme rejuvenation of metallic glasses by shock compression. Science Advances, 2019. 5(8): eaaw6249.

[658]  Priezjev, N.V., Accelerated rejuvenation in metallic glasses subjected to elastostatic compression along alternating directions. Journal of Non-Crystalline Solids, 2021. 556: 120562.

[659]  Guo, W., et al., Various rejuvenation behaviors of Zr-based metallic glass by cryogenic cycling treatment with different casting temperatures. Nanoscale Research Letters, 2018. 13(1): 398.

[660]  Bian, X.L., et al., Atomic origin for rejuvenation of a Zr-based metallic glass at cryogenic temperature. Journal of Alloys and Compounds, 2017. 718: 254–259.

[661]  Das, A., E.M. Dufresne and R. Maaß, Structural dynamics and rejuvenation during cryogenic cycling in a Zr-based metallic glass. Acta Materialia, 2020. 196: 723–732.

[662]  Guo, W., et al., Unconspicuous rejuvenation of a Pd-based metallic glass upon deep cryogenic cycling treatment. Materials Science and Engineering: A, 2019. 759: 59–64.

[663]  Guo, W., et al., Rejuvenation of Zr-based bulk metallic glass matrix composite upon deep cryogenic cycling. Materials Letters, 2019. 247: 135–138.

[664]  Guo, W., et al., Varying the treating conditions to rejuvenate metallic glass by deep cryogenic cycling treatment. Journal of Alloys and Compounds, 2020. 819: 152997.

[665]  Guo, W., R. Yamada and J. Saida, Rejuvenation and plasticization of metallic glass by deep cryogenic cycling treatment. Intermetallics, 2018. 93: 141–147.

[666]  Meylan, C.M., et al., Stimulation of shear-transformation zones in metallic glasses by cryogenic thermal cycling. Journal of Non-Crystalline Solids, 2020. 548: 120299.

[667]  Ri, M.C., et al., Microstructure change in Fe-based metallic glass and nanocrystalline alloy induced by liquid nitrogen treatment. Journal of Materials Science & Technology, 2021. 69: 1–6.

[668]  Ri, M.C., et al., Determination of internal residual strain induced by cryogenic thermal cycling in partially-crystalized Fe-based metallic glasses. Journal of Alloys and Compounds, 2019. 781: 357–361.

[669]  Samavatian, M., et al., Extra rejuvenation of Zr55Cu30Al10Ni5 bulk metallic glass using elastostatic loading and cryothermal treatment interaction. Journal of Non-Crystalline Solids, 2019. 506: 39–45.

[670]  Meylan, C.M., J. Orava and A.L. Greer, Rejuvenation through plastic deformation of a La-based metallic glass measured by fast-scanning calorimetry. Journal of Non-Crystalline Solids: X, 2020. 8: 100051.

[671]  Pan, J., et al., Strain-hardening and suppression of shear-banding in rejuvenated bulk metallic glass. Nature, 2020. 578(7796): 559–562.

[672]  Busch, R., J. Schroers and W.H. Wang, Thermodynamics and Kinetics of Bulk Metallic Glass. MRS Bulletin, 2007. 32(8): 620–623.

[673]  Chen, M., A brief overview of bulk metallic glasses. NPG Asia Mater, 2011. 3.

[674]  Zhu, Y., et al., Development of one-dimensional periodic packing in metallic glass spheres. Scripta Materialia, 2020. 177: 132–136.

[675]  Lou, H., et al., Two-way tuning of structural order in metallic glasses. Nature Communications, 2020. 11(1): 314.

[676]  Zhu, F., et al., Correlation between local structure order and spatial heterogeneity in a metallic glass. Physical Review Letters, 2017. 119(21): 215501.

[677]  Hirata, A., et al., Direct observation of local atomic order in a metallic glass. Nature Materials, 2011. 10(1): 28–33.

[678]  Miracle, D.B., The efficient cluster packing model – An atomic structural model for metallic glasses. Acta Materialia, 2006. 54(16): 4317–4336.

[679]  Miracle, D.B., A structural model for metallic glasses. Nature Materials, 2004. 3(10): 697–702.

[680]  Yavari, A.R., A new order for metallic glasses. Nature, 2006. 439(7075): 405–406.

[681]  Ma, E., Tuning order in disorder. Nature Materials, 2015. 14(6): 547–552.

[682]  Michalik, Š., et al., Short range order and crystallization of Cu–Hf metallic glasses. Journal of Alloys and Compounds, 2021. 853: 156775.

[683]  Schaafsma, A.S., et al., Short Range Order of Metallic Glasses. 1980. 41. http://dx.doi.org/10.1051/jphyscol:1980863,.

[684]  Hayes, T.M., et al., Short-range order in metallic glasses. Physical Review Letters, 1978. 40(19): 1282–1285.

[685]  Hirotsu, Y., T. Ohkubo and M. Matsushita, Study of amorphous alloy structures with medium range atomic ordering. Microscopy Research and Technique, 1998. 40(4): 284–312.

[686]  Cowley, J.M., Electron nanodiffraction methods for measuring medium-range order. Ultramicroscopy, 2002. 90(2): 197–206.

[687]  Xu, Y., et al., Short-to-medium-range order and atomic packing in Zr48Cu36Ag8Al8 bulk metallic glass. Metals, 2016. 6(10).

[688]  Wen, J., et al., Distinguishing medium-range order in metallic glasses using fluctuation electron microscopy: A theoretical study using atomic models. Journal of Applied Physics, 2009. 105(4): 043519.

[689]  Ma, D., A.D. Stoica and X.L. Wang, Power-law scaling and fractal nature of medium-range order in metallic glasses. Nature Materials, 2009. 8(1): 30–34.

[690]  Long, G.G., et al., Highly ordered noncrystalline metallic phase. Physical Review Letters, 2013. 111(1): 015502.

[691]  Zeng, Q., et al., Long-range topological order in metallic glass. Science, 2011. 332(6036): 1404.

[692]  Muley, S.V., et al., Varying kinetic stability, icosahedral ordering, and mechanical properties of a model Zr-Cu-Al metallic glass by sputtering. Physical Review Materials, 2021. 5(3): 033602.

[693]  Hirata, A., et al., Geometric frustration of icosahedron in metallic glasses. Science, 2013. 341(6144): 376.

[694]  Ding, J., M. Asta and R.O. Ritchie, On the question of fractal packing structure in metallic glasses. Proceedings of the National Academy of Sciences, 2017. 114(32): 8458.

[695]  James, P.F., Liquid-phase separation in glass-forming systems. Journal of Materials Science, 1975. 10(10): 1802–1825.

[696]  Alder, B.J. and T.E. Wainwright, Phase transition for a hard sphere system. The Journal of Chemical Physics, 1957. 27(5): 1208–1209.

[697] Calvo-Dahlborg, M., et al., Structural study of a phase transition in a NiP metallic glass. Materials Science and Engineering, 1997. 226.

[698] Chou, H.S., et al., Time-dependent mechanical behavior of phase-separated ZrCuTiTa thin films under nanoindentation. Intermetallics, 2012. 27: 26–30.

[699] Feng, S., et al., A general and transferable deep learning framework for predicting phase formation in materials. npj Computational Materials, 2021. 7(1): 10.

[700] Straumal, B.B., et al., Phase transitions in Cu-based alloys under high pressure torsion. Journal of Alloys and Compounds, 2017. 707: 20–26.

[701] Zheng, J., et al., Temperature-resistance sensor arrays for combinatorial study of phase transitions in shape memory alloys and metallic glasses. Scripta Materialia, 2019. 168: 144–148.

[702] Xie, S., X. Tu and J.J. Kruzic, Inducing strain hardening in a Zr-based bulk metallic glass via cobalt mediated phase separations. Journal of Alloys and Compounds, 2018. 735.

[703] Sarlar, K. and I. Kucuk, Phase separation and glass forming ability of (Fe0.72Mo0.04B0.24) 100-xGdx (x=4, 8) bulk metallic glasses. Journal of Non-Crystalline Solids, 2016. 447: 198–201.

[704] Cadien, A., et al., First-order liquid-liquid phase transition in cerium. Physical Review Letters, 2013. 110(12): 125503.

[705] Cao, C.R., et al., Liquid-like behaviours of metallic glassy nanoparticles at room temperature. Nature Communications, 2019. 10.

[706] Chen, E.-Y., et al., Glass-forming ability correlated with the liquid-liquid transition in Pd42.5Ni42.5P15 alloy. Scripta Materialia, 2021. 193: 117–121.

[707] Cohen, M.H. and G.S. Grest, Liquid-glass transition a free-volume approach. Physical Review B, 1979. 20.

[708] Albe, K., Y. Ritter and D. Şopu, Enhancing the plasticity of metallic glasses: Shear band formation, nanocomposites and nanoglasses investigated by molecular dynamics simulations. Mechanics of Materials, 2013. 67: 94–103.

[709] Binkowski, I., et al., Shear band relaxation in a deformed bulk metallic glass. Acta Materialia, 2016. 109: 330–340.

[710] Park, E.S., Understanding of the shear bands in amorphous metals. Applied Microscopy, 2015. 45.

[711] Wang, D.P., et al., Mutual interaction of shear bands in metallic glasses. Intermetallics, 2017. 85: 48–53.

[712] Perepezko, J.H., et al., Nucleation of shear bands in amorphous alloys. Proceedings of the National Academy of Sciences, 2014. 111(11): 3938.

[713] Yan, Z., et al., Localized crystallization in shear bands of a metallic glass. Scientific Reports, 2016. 6(1): 19358.

[714] Qu, R.T., et al., Progressive shear band propagation in metallic glasses under compression. Acta Materialia, 2015. 91.

[715] Song, H.Y., et al., Effect of crystal phase on shear bands initiation and propagation behavior in metallic glass matrix composites. Computational Materials Science, 2018. 150: 42–46.

[716] Maaß, R. and J.F. Löffler, Shear-band dynamics in metallic glasses. Advanced Functional Materials, 2015. 25(16): 2353–2368.

[717] Klaumünzer, D., et al., Temperature-dependent shear band dynamics in a Zr-based bulk metallic glass. Applied Physics Letters, 2010. 96(6): 061901.

[718] Guduru, P.R., A.J. Rosakis and G. Ravichandran, Dynamic shear bands: an investigation using high speed optical and infrared diagnostics. Mechanics of Materials, 2001. 33(7): 371–402.

[719]    Qu, R.T., et al., Shear band fracture in metallic glass: Hot or cold?. Scripta Materialia, 2019.
         162: 136–140.
[720]    Wang, X.D., et al., Shear band-mediated fatigue cracking mechanism of metallic glass at
         high stress level. Materials Science and Engineering: A, 2015. 627: 336–339.
[721]    Zhong, C., et al., Relationship of deformation mode with strain-dependent shear
         transformation zone size in Cu-Zr metallic glasses using molecular dynamics simulations.
         Journal of Non-Crystalline Solids, 2017. 469: 45–50.
[722]    Maaß, R., et al., Single shear-band plasticity in a bulk metallic glass at cryogenic
         temperatures. Scripta Materialia, 2012. 66(5): 231–234.
[723]    Meduri, C., et al., Effect of temperature on shear bands and bending plasticity of metallic
         glasses. Journal of Alloys and Compounds, 2018. 732: 922–927.
[724]    Narayan, R.L., D. Raut and U. Ramamurty, A quantitative connection between shear band
         mediated plasticity and fracture initiation toughness of metallic glasses. Acta Materialia,
         2018. 150: 69–77.
[725]    Yang, G.N., et al., Understanding the effects of Poisson's ratio on the shear band behavior
         and plasticity of metallic glasses. Journal of Materials Science, 2017. 52(11): 6789–6799.
[726]    Yang, G.N., et al., The multiple shear bands and plasticity in metallic glasses: A possible
         origin from stress redistribution. Journal of Alloys and Compounds, 2017. 695: 3457–3466.
[727]    Wang, Z., Serration Behavior in Pd77.5Cu6Si16.5 Alloy. Metals, 2016. 6(8).
[728]    Xie, X., et al., Origin of serrated flow in bulk metallic glasses. Journal of the Mechanics and
         Physics of Solids, 2019. 124: 634–642.
[729]    Chen, H.M., et al., Flow serration and shear-band propagation in bulk metallic glasses.
         Applied Physics Letters, 2009. 94(14): 141914.
[730]    Ren, J.L., et al., Dynamics of serrated flow in a bulk metallic glass. AIP Advances, 2011. 1(3):
         032158.
[731]    Cheng, L., et al., Serrated flow behaviors of a Zr-based bulk metallic glass by
         nanoindentation. Journal of Applied Physics, 2014. 115(8): 084907.
[732]    Fan, Z., et al., Strategies to tailor serrated flows in metallic glasses. Journal of Materials
         Research, 2019. 34(9): 1595–1607.
[733]    Maaß, R., Beyond serrated flow in bulk metallic glasses: what comes next?. Metallurgical
         and Materials Transactions A, 2020. 51(11): 5597–5605.
[734]    Haichao Sun, Z.N., J. Ren, W. Liang, Y. Huang, J. Sun, X. Xue and G. Wang, Serration and
         shear avalanches in a ZrCu based bulk metallic glass composite in different loading
         methods. Journal of Materials Science & Technology, 2019. 35(9): 2079–2085.
[735]    Antonaglia, J., et al., Tuned critical avalanche scaling in bulk metallic glasses. Scientific
         Reports, 2014. 4(1): 4382.
[736]    Kosiba, K., et al., Modulating heterogeneity and plasticity in bulk metallic glasses: Role of
         interfaces on shear banding. International Journal of Plasticity, 2019. 119: 156–170.
[737]    Huang, E., P.K. Liaw and G. Editors, High-temperature materials for structural applications:
         new perspectives on high-entropy alloys, bulk metallic glasses, and nanomaterials. MRS
         Bull, 2019. 44.
[738]    Bizhanova, G., et al., Development and crystallization kinetics of novel near-equiatomic
         high-entropy bulk metallic glasses. Journal of Alloys and Compounds, 2019. 779: 474–486.
[739]    Chen, Y., Z.-W. Dai and J.-Z. Jiang, High entropy metallic glasses: Glass formation,
         crystallization and properties. Journal of Alloys and Compounds, 2021. 866: 158852.
[740]    Inoue, A., et al., Formation, structure and properties of pseudo-high entropy clustered bulk
         metallic glasses. Journal of Alloys and Compounds, 2020. 820: 153164.
[741]    Zhao, S.F., et al., Pseudo-quinary Ti20Zr20Hf20Be20(Cu20-xNix) high entropy bulk metallic
         glasses with large glass forming ability. Materials & Design, 2015. 87: 625–631.

[742]   Ding, H.Y., et al., A senary TiZrHfCuNiBe high entropy bulk metallic glass with large glass-forming ability. Materials Letters, 2014. 125: 151–153.

[743]   Gong, P., J. Jin and L. Deng, Room temperature nanoindentation creep behavior of TiZrHfBeCu(Ni) high entropy bulk metallic glasses. Materials Science and Engineering, 2017. 688.

[744]   Gong, P., et al., Research on nano-scratching behavior of TiZrHfBeCu(Ni) high entropy bulk metallic glasses. Journal of Alloys and Compounds, 2020. 817: 153240.

[745]   Zhang, M., et al., Oxidation behavior of a Ti16.7Zr16.7Hf16.7Cu16.7Ni16.7Be16.7 high-entropy bulk metallic glass. Materials Letters, 2019. 236.

[746]   Li, M., et al., Minor Cr alloyed Fe–Co–Ni–P–B high entropy bulk metallic glass with excellent mechanical properties. Materials Science and Engineering: A, 2021. 805: 140542.

[747]   Li, N., et al., Fe-based metallic glass reinforced FeCoCrNiMn high entropy alloy through selective laser melting. Journal of Alloys and Compounds, 2020. 822: 153695.

[748]   Pang, C.M., et al., Effect of Yttrium addition on magnetocaloric properties of Gd-Co-Al-Ho high entropy metallic glasses. Journal of Non-Crystalline Solids, 2020. 549: 120354.

[749]   Zhao, Y., et al., The microalloying effect of Ce on the mechanical properties of medium entropy bulk metallic glass composites. Crystals, 2019. 9(9).

[750]   Zhao, S.F., et al., A quinary Ti–Zr–Hf–Be–Cu high entropy bulk metallic glass with a critical size of 12 mm. Intermetallics, 2015. 61: 47–50.

[751]   Wang, X., et al., Thermoplastic micro-formability of TiZrHfNiCuBe high entropy metallic glass. Journal of Materials Science & Technology, 2018. 34(11): 2006–2013.

[752]   Chen, N., D.V. Louzguine-Luzgin and K. Yao, A new class of non-crystalline materials: Nanogranular metallic glasses. Journal of Alloys and Compounds, 2017. 707: 371–378.

[753]   Gleiter, H., Nanoglasses: a new kind of noncrystalline materials. The Beilstein Journal of Nanotechnology, 2013. 4.

[754]   Adibi, S., et al., Composition and grain size effects on the structural and mechanical properties of CuZr nanoglasses. Journal of Applied Physics, 2014. 116(4): 043522.

[755]   Adjaoud, O. and K. Albe, Interfaces and interphases in nanoglasses: Surface segregation effects and their implications on structural properties. Acta Materialia, 2016. 113: 284–292.

[756]   Ghafari, M., Structural investigations of interfaces in Fe90Sc10 nanoglasses using high-energy x-ray diffraction. Applied Physics Letters, 2012. 100.

[757]   Adjaoud, O. and K. Albe, Microstructure formation of metallic nanoglasses: Insights from molecular dynamics simulations. Acta Materialia, 2018. 145: 322–330.

[758]   Adjaoud, O. and K. Albe, Influence of microstructural features on the plastic deformation behavior of metallic nanoglasses. Acta Materialia, 2019. 168: 393–400.

[759]   Cheng, B. and J.R. Trelewicz, Interfacial plasticity governs strain delocalization in metallic nanoglasses. Journal of Materials Research, 2019. 34(13): 2325–2336.

[760]   Kalcher, C., et al., Reinforcement of nanoglasses by interface strengthening. Scripta Materialia, 2017. 141: 115–119.

[761]   Ritter, Y., et al., Structure, stability and mechanical properties of internal interfaces in Cu64Zr36 nanoglasses studied by MD simulations. Acta Materialia, 2011. 59(17): 6588–6593.

[762]   Sha, Z.-D., et al., Strong and ductile nanolaminate composites combining metallic glasses and nanoglasses. International Journal of Plasticity, 2017. 90: 231–241.

[763]   Wang, X., et al., Sample size effects on strength and deformation mechanism of Sc75Fe25 nanoglass and metallic glass. Scripta Materialia, 2016. 116: 95–99.

[764]   Śniadecki, Z., et al., Nanoscale morphology of Ni50Ti45Cu5 nanoglass. Materials Characterization, 2016. 113: 26–33.

[765]   Wang, C., et al., Atomic structure of Fe90Sc10 glassy nanoparticles and nanoglasses. Scripta Materialia, 2017. 139: 9–12.

[766] Wang, C., et al., Tuning the Curie temperature of Fe90Sc10 nanoglasses by varying the volume fraction and the composition of the interfaces. Scripta Materialia, 2019. 159: 109–112.

[767] Cao, Q.P., et al., Effect of Nb substitution for Cu on glass formation and corrosion behavior of ZrCuAgAlBe bulk metallic glass. Journal of Alloys and Compounds, 2016. 683: 22–31.

[768] Fornell, J., et al., Improved plasticity and corrosion behavior in Ti–Zr–Cu–Pd metallic glass with minor additions of Nb: An alloy composition intended for biomedical applications. Materials Science and Engineering: A, 2013. 559: 159–164.

[769] Gu, J.-L., et al., Effects of Cu addition on the glass forming ability and corrosion resistance of Ti-Zr-Be-Ni alloys. Journal of Alloys and Compounds, 2017. 725: 573–579.

[770] Guo, Y., et al., Ni- and Cu-free Ti-based metallic glasses with potential biomedical application. Intermetallics, 2015. 63: 86–96.

[771] Guo, Y., et al., On the ternary eutectic reaction in the Fe60Cr8Nb8B24 quaternary alloy. Journal of Alloys and Compounds, 2017. 707: 281–286.

[772] Jia, C.G., et al., Tailoring the corrosion behavior of Fe-based metallic glasses through inducing Nb-triggered netlike structure. Corrosion Science, 2019. 147: 94–107.

[773] Li, C.-L., Effect of nitrogen content on the corrosion resistance of Zr- Ni-Al-Si thin film metallic glass. International Journal of Electrochemical Science, 2017. 12: 12074–12083.

[774] Lin, H.C., et al., Designing a toxic-element-free Ti-based amorphous alloy with remarkable supercooled liquid region for biomedical application. Intermetallics, 2014. 55: 22–27.

[775] Xu, D.D., et al., Effects of Cr addition on thermal stability, soft magnetic properties and corrosion resistance of FeSiB amorphous alloys. Corrosion Science, 2018. 138: 20–27.

[776] Gebert, A., P.F. Gostin and L. Schultz, Effect of surface finishing of a Zr-based bulk metallic glass on its corrosion behaviour. Corrosion Science, 2010. 52(5): 1711–1720.

[777] Smugeresky, J., et al., Laser engineered net shaping(LENS) process: optimization of surface finish and microstructural properties. Advances in Powder Metallurgy and Particulate Materials–1997, 1997. 3: 21.

[778] Deng, L., et al., Mechanical performance and corrosion behaviour of Zr-based bulk metallic glass produced by selective laser melting. Materials & Design, 2020. 189: 108532.

[779] Wang, Q.-Y., et al., Effects of laser re-melting and annealing on microstructure, mechanical property and corrosion resistance of Fe-based amorphous/crystalline composite coating. Materials Characterization, 2017. 127: 239–247.

[780] Yoshioka, H., et al., Laser-processed corrosion-resistant amorphous Ni–Cr–P–B surface alloys on a mild steel. Corrosion Science, 1987. 27(9): 981–995.

[781] Ibrahim, M.Z., et al., Characterization and hardness enhancement of amorphous Fe-based metallic glass laser cladded on nickel-free stainless steel for biomedical implant application. Materials Chemistry and Physics, 2019. 235: 121745.

[782] Qiu, X.-W. and C.-G. Liu, Microstructure and properties of Al2CrFeCoCuTiNix high-entropy alloys prepared by laser cladding. Journal of Alloys and Compounds, 2013. 553: 216–220.

[783] Zhang, C., et al., 3D printing of Zr-based bulk metallic glasses and components for potential biomedical applications. Journal of Alloys and Compounds, 2019. 790: 963–973.

[784] Chu, J.P., et al., Fabrication and characterizations of thin film metallic glasses: Antibacterial property and durability study for medical application. Thin Solid Films, 2014. 561: 102–107.

[785] Chen, L.T., Microstructure, mechanical and anti-corrosion property evaluation of iron-based thin film metallic glasses. Surface and Coatings Technology, 2014. 260.

[786] Huang, Y., et al., Synthesis of Fe–Cr–Mo–C–B amorphous coating with high corrosion resistance. Materials Letters, 2012. 89: 229–232.

[787] Li, Z., C. Zhang and L. Liu, Wear behavior and corrosion properties of Fe-based thin film metallic glasses. Journal of Alloys and Compounds, 2015. 650.

[788]    Lim, K.R., et al., Stress-driven crystallization of amorphous metal oxide film by glass
         transition in metallic glass. Corrosion Science, 2016. 104: 98–102.

[789]    Xi, Y., D. Liu and D. Han, Improvement of corrosion and wear resistances of AISI 420
         martensitic stainless steel using plasma nitriding at low temperature. Surface and Coatings
         Technology, 2008. 202.

[790]    Kawashima, A., et al., Pitting Corrosion of Amorphous Ni–Zr Alloys in Chloride Ion
         Containing Sulfuric Acid Solutions. Materials Transactions, JIM, 1997. 38(5): 443–450.

[791]    Mudali, U.K., et al., Pitting corrosion of bulk glass-forming zirconium-based alloys. Journal
         of alloys and compounds, 2004. 377(1–2): 290–297.

[792]    Wang, Y.M., et al., Why does pitting preferentially occur on shear bands in bulk metallic
         glasses?. Intermetallics, 2013. 42: 107–111.

[793]    Zhou, M., et al., Improved mechanical properties and pitting corrosion resistance of
         Zr65Cu17.5Fe10Al7.5 bulk metallic glass by isothermal annealing. Journal of Non-Crystalline
         Solids, 2016. 452: 50–56.

[794]    Hua, N., et al., Tribological behavior of a Ni-free Zr-based bulk metallic glass with potential
         for biomedical applications. Materials Science and Engineering: C, 2016. 66: 268–277.

[795]    Hua, N., et al., Effects of noble elements on the glass-forming ability, mechanical property,
         electrochemical behavior and tribocorrosion resistance of Ni- and Cu-free Zr-Al-Co bulk
         metallic glass. Journal of Alloys and Compounds, 2017. 725: 403–414.

[796]    Jiang, X., et al., Tribological behaviors of self–mated Cu36Zr48Ag8Al8 bulk metallic glass
         under H2SO4 conditions with different concentrations. Tribology International, 2019. 136:
         395–403.

[797]    Geissler, D., M. Uhlemann and A. Gebert, Catastrophic stress corrosion failure of Zr-base
         bulk metallic glass through hydrogen embrittlement. Corrosion Science, 2019. 159: 108057.

[798]    Kawashima, A., et al., A Ni- and Cu-free Zr-based bulk metallic glass with excellent
         resistance to stress corrosion cracking in simulated body fluids. Materials Science and
         Engineering:
         A, 2012. 542: 140–146.

[799]    Nakai, Y. and Y. Yoshioka, Stress corrosion and corrosion fatigue crack growth of Zr-based
         bulk metallic glass in aqueous solutions. Metallurgical and Materials Transactions A, 2010.
         41(7): 1792–1798.

[800]    Scully, J.R., A. Gebert and J.H. Payer, Corrosion and related mechanical properties of bulk
         metallic glasses. Journal of Materials Research, 2007. 22(2): 302–313.

[801]    Guo, R.Q., et al., Corrosion and wear resistance of a Fe-based amorphous coating in
         underground environment. Intermetallics, 2012. 30: 94–99.

[802]    Han, K., et al., Antimicrobial Zr-based bulk metallic glasses for surgical devices
         applications. Journal of Non-Crystalline Solids, 2021. 564: 120827.

[803]    Prabhu, Y., et al., Cu-Zr-Ti-Al metallic glass: Thermodynamic prediction, synthesis and
         biocorrosion studies. Physica B: Condensed Matter, 2021. 609: 412918.

[804]    Al-Tamimi, A.A., et al., Topology optimised metallic bone plates produced by electron beam
         melting: a mechanical and biological study. The International Journal of Advanced
         Manufacturing Technology, 2019. 104(1): 195–210.

[805]    Han, K., et al., Low-cost Zr-based bulk metallic glasses for biomedical devices applications.
         Journal of Non-Crystalline Solids, 2019. 520: 119442.

[806]    Jabed, A., et al., Property optimization of Zr-Ti-X (X = Ag, Al) metallic glass via combinatorial
         development aimed at prospective biomedical application. Surface and Coatings
         Technology, 2019. 372: 278–287.

[807]    Khan, M.M., K.M. Deen and W. Haider, Combinatorial development and assessment of a
         Zr-based metallic glass for prospective biomedical applications. Journal of Non-Crystalline
         Solids, 2019. 523: 119544.
[808]    Khan, M.M., I. Shabib and W. Haider, A combinatorially developed Zr-Ti-Fe-Al metallic glass
         with outstanding corrosion resistance for implantable medical devices. Scripta Materialia,
         2019. 162.
[809]    Li, J., et al., Zr61Ti2Cu25Al12 metallic glass for potential use in dental implants:
         Biocompatibility assessment by in vitro cellular responses. Materials Science and
         Engineering: C, 2013. 33(4): 2113–2121.
[810]    Lin, C.H., et al., Rapid screening of potential metallic glasses for biomedical applications.
         Materials Science and Engineering: C, 2013. 33(8): 4520–4526.
[811]    Pang, S., et al., New Ti-based Ti–Cu–Zr–Fe–Sn–Si–Ag bulk metallic glass for biomedical
         applications. Journal of Alloys and Compounds, 2015. 625: 323–327.
[812]    Perumal, G., et al., Enhancement in bio-corrosion resistance of metallic glass by severe
         surface deformation. Applied Surface Science, 2019. 487: 1096–1103.
[813]    Sun, Y., et al., In vitro and in vivo biocompatibility of an Ag-bearing Zr-based bulk metallic
         glass for potential medical use. Journal of Non-Crystalline Solids, 2015. 419: 82–91.
[814]    Thanka Rajan, S., A. Bendavid and B. Subramanian, Cytocompatibility assessment of
         Ti-Nb-Zr-Si thin film metallic glasses with enhanced osteoblast differentiation for
         biomedical applications. Colloids and Surfaces B: Biointerfaces, 2019. 173: 109–120.
[815]    Wang, C., et al., Ti-Cu-Zr-Fe-Sn-Si-Ag-Pd bulk metallic glasses with potential for biomedical
         applications. Metallurgical and Materials Transactions A, 2021. 52(5): 1559–1567.
[816]    Wang, G., et al., A new TiCuHfSi bulk metallic glass with potential for biomedical
         applications. Materials & Design (1980–2015), 2014. 54: 251–255.
[817]    Wang, Y.B., et al., In vitro and in vivo studies on Ti-based bulk metallic glass as potential
         dental implant material. Materials Science and Engineering: C, 2013. 33(6): 3489–3497.
[818]    Wang, Y.B., et al., Corrosion performances in simulated body fluids and cytotoxicity
         evaluation of Fe-based bulk metallic glasses. Materials Science and Engineering: C, 2012.
         32(3): 599–606.
[819]    Zhang, X.L., G. Chen and T. Bauer, Mg-based bulk metallic glass composite with high
         bio-corrosion resistance and excellent mechanical properties. Intermetallics, 2012. 29:
         56–60.
[820]    Aliyu, A.A.A., et al., Characterization, adhesion strength and in-vitro cytotoxicity
         investigation of hydroxyapatite coating synthesized on Zr-based BMG by electro discharge
         process.
         Surf Coat Technol, 2019. 370.
[821]    Huang, L., et al., A Zr-based bulk metallic glass for future stent applications: Materials
         properties, finite element modeling, and in vitro human vascular cell response. Acta
         Biomaterialia, 2015. 25: 356–368.
[822]    Escher, B., et al., Stability of the B2 CuZr phase in Cu-Zr-Al-Sc bulk metallic glass matrix
         composites. Journal of Alloys and Compounds, 2019. 790: 657–665.
[823]    Li, F.X., et al., Anisotropic elastic properties and phase stability of B2 and B19 CuZr
         structures under hydrostatic pressure. Intermetallics, 2018. 98: 60–68.
[824]    Amigo, N., M. Sepúlveda-Macías and G. Gutiérrez, Martensitic transformation to monoclinic
         phase in bulk B2–CuZr. Intermetallics, 2017. 91: 16–21.
[825]    González, S., et al., Drastic influence of minor Fe or Co additions on the glass forming
         ability, martensitic transformations and mechanical properties of shape memory Zr–Cu–Al
         bulk metallic glass composites. Science and Technology of Advanced Materials, 2014.
         15(3): 035015.

[826] Hong, S.H., et al., Ultrafine shape memory alloys synthesized using a metastable metallic glass precursor with polymorphic crystallization. Applied Materials Today, 2021. 22: 100961.

[827] Song, K.K., et al., Strategy for pinpointing the formation of B2 CuZr in metastable CuZr-based shape memory alloys. Acta Mater, 2011. 59.

[828] Kozachkov, H., et al., Effect of cooling rate on the volume fraction of B2 phases in a CuZrAlCo metallic glass matrix composite. Intermetallics, 2013. 39: 89–93.

[829] Kosiba, K., et al., Transient nucleation and microstructural design in flash-annealed bulk metallic glasses. Acta Materialia, 2017. 127: 416–425.

[830] Akihisa, I., et al., Aluminum-based amorphous alloys with tensile strength above 980 MPa (100 kg/mm2). Japanese Journal of Applied Physics, 1988. 27(4A): L479.

[831] Inoue, A., Bulk amorphous and nanocrystalline alloys with high functional properties. Materials Science and Engineering: A, 2001. 304–306: 1–10.

[832] Qian, M., Metal powder for additive manufacturing. JOM, 2015. 67(3): 536–537.

[833] Gibson, I., W.D. Rosen and B. Stucker, Medical Applications for Additive Manufacture, in Additive Manufacturing Technologies: Rapid Prototyping to Direct Digital Manufacturing. 2010, Springer US: Boston, MA. 400–414.

[834] Kelly, P.M. and M.-X. Zhang, Edge-to-edge matching – the fundamentals. Metallurgical and Materials Transactions A, 2006. 37(3): 833–839.

[835] Kelly, P. and M.-X. Zhang, Edge-to-edge matching-a new approach to the morphology and crystallography of precipitates. Materials Forum, 1999.

[836] Zhang, M.X. and P.M. Kelly, Edge-to-edge matching and its applications: Part I. Application to the simple HCP/BCC system. Acta Materialia, 2005. 53(4): 1073–1084.

[837] Zhang, M.X. and P.M. Kelly, Edge-to-edge matching model for predicting orientation relationships and habit planes – the improvements. Scripta Materialia, 2005. 52(10): 963–968.

[838] Pinkerton, A.J., [INVITED] Lasers in additive manufacturing. Optics & Laser Technology, 2016. 78(Part A): 25–32.

[839] Rombouts, M., Selective laser sintering/melting of iron-based powders (Selectief laser sinteren/smelten van ijzergebaseerde poeders). 2006,

[840] Badrossamay, M. and T.H.C. Childs, Further studies in selective laser melting of stainless and tool steel powders. International Journal of Machine Tools and Manufacture, 2007. 47(5): 779–784.

[841] Kempen, K., et al., Mechanical properties of AlSi10Mg produced by selective laser melting. Physics Procedia, 2012. 39: 439–446.

[842] Taltavull, C., et al., Selective laser surface melting of a magnesium-aluminium alloy. Materials Letters, 2012. 85: 98–101.

[843] Leuders, S., et al., On the mechanical behaviour of titanium alloy TiAl6V4 manufactured by selective laser melting: Fatigue resistance and crack growth performance. International Journal of Fatigue, 2013. 48: 300–307.

[844] Vandenbroucke, B. and J.P. Kruth, Selective laser melting of biocompatible metals for rapid manufacturing of medical parts. Rapid Prototyping Journal, 2007. 13(4): 196–203.

[845] Warnke, P.H., et al., Rapid prototyping: porous titanium alloy scaffolds produced by selective laser melting for bone tissue engineering. Tissue Engineering Part C: Methods, 2008. 15(2): 115–124.

[846] Tang, Y., et al., Direct laser sintering of a copper-based alloy for creating three-dimensional metal parts. Journal of Materials Processing Technology, 2003. 140(1–3): 368–372.

[847] Yves-Christian, H., et al., Net shaped high performance oxide ceramic parts by selective laser melting. Physics Procedia, 2010. 5: 587–594.

[848]   Shishkovsky, I., et al., Alumina–zirconium ceramics synthesis by selective laser sintering/melting. Applied Surface Science, 2007. 254(4): 966–970.

[849]   Luo, J., H. Pan and E.C. Kinzel, Additive manufacturing of glass. Journal of Manufacturing Science and Engineering, 2014. 136(6): 061024.

[850]   Quadrini, F. and L. Santo, Selective laser sintering of resin-coated sands – Part I: the laser-material interaction. Journal of Manufacturing Science and Engineering, 2008. 131(1): 011004–011004.

[851]   Niino, Toshiki & Yamada, Hidenori. (2009). Fabrication of Transparent Parts by Laser Sintering Process. Journal of the Japan Society for Precision Engineering. 75. 1454–1458. 10.2493/jjspe.75.1454.

[852]   Mazzoli, A., G. Moriconi and M.G. Pauri, Characterization of an aluminum-filled polyamide powder for applications in selective laser sintering. Materials & Design, 2007. 28(3): 993–1000.

[853]   Kandis, M. and T.L. Bergman, A Simulation-based correlation of the density and thermal conductivity of objects produced by laser sintering of polymer powders. Journal of Manufacturing Science and Engineering, 1999. 122(3): 439–444.

[854]   Childs, T., et al. Simulation and experimental verification of crystalline polymer and direct metal selective laser sintering, in Proc. SFF Symp., Austin. 2000.

[855]   Berzins, M., T.H.C. Childs and G.R. Ryder, The selective laser sintering of polycarbonate. CIRP Annals – Manufacturing Technology, 1996. 45(1): 187–190.

[856]   Hui, Z., et al., Microstructure and age characterization of Cu–15Ni–8Sn alloy coatings by laser cladding. Applied Surface Science, 2010. 256(20): 5837–5842.

[857]   Zhang, H., Y. Pan and Y. He, Effects of annealing on the microstructure and properties of 6FeNiCoCrAlTiSi high-entropy alloy coating prepared by laser cladding. Journal of Thermal Spray Technology, 2011. 20(5): 1049–1055.

[858]   Zhang, H., et al., Microstructure and properties of 6FeNiCoSiCrAlTi high-entropy alloy coating prepared by laser cladding. Applied Surface Science, 2011. 257(6): 2259–2263.

[859]   Zhang, H., Y. Pan and Y.-Z. He, Synthesis and characterization of FeCoNiCrCu high-entropy alloy coating by laser cladding. Materials & Design, 2011. 32(4): 1910–1915.

[860]   Schmidt, M., et al., Lasers in Manufacturing 2011 – Proceedings of the Sixth International WLT Conference on Lasers in Manufacturing the Property Research on High-entropy Alloy AlxFeCoNiCuCr Coating by Laser Cladding. Physics Procedia, 2011. 12: p. 303–312.

[861]   Wu, W., et al., Phase evolution and properties of Al2CrFeNiMox high-entropy alloys coatings by laser cladding. Journal of Thermal Spray Technology, 2015. 24(7): 1333–1340.

[862]   Kim, J., et al., Phase evolution in Cu54Ni6Zr22Ti18 bulk metallic glass Nd:YAG laser weld. Materials Science and Engineering: A, 2006. 434(1–2): 194–201.

[863]   Glardon, R., et al., Influence of Nd:YAG parameters on the Selective Laser Sintering of Metallic Powders. CIRP Annals – Manufacturing Technology, 2001. 50(1): 133–136.

[864]   Lin, H.-K., et al., Pulsed laser micromachining of Mg–Cu–Gd bulk metallic glass. Optics and Lasers in Engineering, 2012. 50(6): 883–886.

[865]   Quintana, I., et al., Investigation of amorphous and crystalline Ni alloys response to machining with micro-second and pico-second lasers. Applied Surface Science, 2009. 255(13–14): 6641–6646.

[866]   Jia, W., et al., The effect of femtosecond laser micromachining on the surface characteristics and subsurface microstructure of amorphous FeCuNbSiB alloy. Applied Surface Science, 2006. 253(3): 1299–1303.

[867]   Wang, X., et al., Noncrystalline micromachining of amorphous alloys using femtosecond laser pulses. Materials Letters, 2007. 61(21): 4290–4293.

[868]  Jia, W., et al., Heat effects of amorphous FeCuNbSiB alloy ablated with femtosecond laser. Thin Solid Films, 2008. 516(8): 2260–2263.

[869]  Ramil, A., et al., Micromachining of glass by the third harmonic of nanosecond Nd: YVO4 laser. Applied Surface Science, 2009. 255(10): 5557–5560.

[870]  Ma, F., et al., Femtosecond laser-induced concentric ring microstructures on Zr-based metallic glass. Applied Surface Science, 2010. 256(11): 3653–3660.

[871]  Miracle, D., et al., Exploration and development of high entropy alloys for structural applications. Entropy, 2014. 16(1): 494.

[872]  Brif, Y., M. Thomas and I. Todd, The use of high-entropy alloys in additive manufacturing. Scripta Materialia, 2015. 99: 93–96.

[873]  Tang, H.P., et al., Additive manufacturing of a high niobium-containing titanium aluminide alloy by selective electron beam melting. Materials Science and Engineering: A, 2015. 636: 103–107.

[874]  Körner, C., Additive manufacturing of metallic components by selective electron beam melting – a review. International Materials Reviews, 2016. 61(5): 361–377.

[875]  Marion, G., et al., Modelling additive manufacturing processes: from direct metal deposition to selective laser melting. S09a Procédés de mise en forme, 2015.

[876]  Marion, G., et al., A finite element model for the simulation of direct metal deposition.

[877]  Amine, T., J.W. Newkirk and F. Liou, Investigation of effect of process parameters on multilayer builds by direct metal deposition. Applied Thermal Engineering, 2014. 73(1): 500–511.

[878]  Shukla, M. and V. Verma, Finite element simulation and analysis of laser metal deposition.

[879]  Zekovic, S., R. Dwivedi and R. Kovacevic. Thermo-structural finite element analysis of direct laser metal deposited thin-walled structures, in Proceedings SFF Symposium,. Austin, TX. 2005.

[880]  Amine, T., J.W. Newkirk and F. Liou, An investigation of the effect of direct metal deposition parameters on the characteristics of the deposited layers. Case Studies in Thermal Engineering, 2014. 3: 21–34.

[881]  Shiomi, M., et al., Finite element analysis of melting and solidifying processes in laser rapid prototyping of metallic powders. International Journal of Machine Tools and Manufacture, 1999. 39(2): 237–252.

[882]  Liu, Z. and H. Qi, Numerical simulation of transport phenomena for a double-layer laser powder deposition of single-crystal superalloy. Metallurgical and Materials Transactions A, 2014. 45(4): 1903–1915.

[883]  Hebert, R.J., Viewpoint: metallurgical aspects of powder bed metal additive manufacturing. Journal of Materials Science, 2016. 51(3): 1165–1175.

[884]  Kathuria, Y.P., Microstructuring by selective laser sintering of metallic powder. Surface and Coatings Technology, 1999. 116–119: 643–647.

[885]  King, W., et al., Overview of modelling and simulation of metal powder bed fusion process at Lawrence Livermore National Laboratory. Materials Science and Technology, 2015. 31(8): 957–968.

[886]  Kumar, S., Selective laser sintering: A qualitative and objective approach. JOM, 2003. 55 (10): 43–47.

[887]  Lee, Y.S. and W. Zhang, Modeling of heat transfer, fluid flow and solidification microstructure of nickel-base superalloy fabricated by laser powder bed fusion. Additive Manufacturing.

[888]  Matsumoto, M., et al., Finite element analysis of single layer forming on metallic powder bed in rapid prototyping by selective laser processing. International Journal of Machine Tools and Manufacture, 2002. 42(1): 61–67.

[889]   Lee, Y., Simulation of Laser Additive Manufacturing and its Applications. 2015, The Ohio State University.

[890]   Griffith, M.L., et al., Understanding the microstructure and properties of components fabricated by laser engineered net shaping (LENS), in: MRS Proceedings. 2011. 625.

[891]   Matthews, D.T.A., et al., Laser engineered surfaces from glass forming alloy powder precursors: Microstructure and wear. Surface and Coatings Technology, 2009. 203(13): 1833–1843.

[892]   Smith, J., et al., Linking process, structure, property, and performance for metal-based additive manufacturing: computational approaches with experimental support. Computational Mechanics, 2016. 57(4): 583–610.

[893]   Khairallah, S.A., et al., Laser powder-bed fusion additive manufacturing: Physics of complex melt flow and formation mechanisms of pores, spatter, and denudation zones. Acta Materialia, 2016. 108: 36–45.

[894]   Deckard, C.R., Method and Apparatus for Producing Parts by Selective Sintering. 1988, Google Patents.

[895]   Meiners, W., K.D. Wissenbach and A.D. Gasser, Shaped Body Especially Prototype or Replacement Part Production. 1998, Google Patents.

[896]   Van Elsen, M., Complexity of Selective Laser Melting: A New Optimisation Approach. 2007, Citeseer.

[897]   Mani, M., et al., Measurement science needs for real-time control of additive manufacturing powder bed fusion processes. National Institute of Standards and Technology, Gaithersburg, MD, NIST Interagency/Internal Report (NISTIR), 2015. 8036.

[898]   Dass, A. and A. Moridi, State of the art in directed energy deposition: from additive manufacturing to materials design. Coatings, 2019. 9(7).

[899]   Syed, W.U.H., A.J. Pinkerton and L. Li, A comparative study of wire feeding and powder feeding in direct diode laser deposition for rapid prototyping. Applied Surface Science, 2005. 247(1): 268–276.

[900]   Syed, W.U.H., A.J. Pinkerton and L. Li, Simultaneous wire- and powder-feed direct metal deposition: An investigation of the process characteristics and comparison with single-feed methods. Journal of Laser Applications, 2006. 18(1): 65–72.

[901]   Khorasani, A., et al., A review of technological improvements in laser-based powder bed fusion of metal printers. The International Journal of Advanced Manufacturing Technology, 2020. 108(1): 191–209.

[902]   Raghavan, A., et al., Heat transfer and fluid flow in additive manufacturing. Journal of Laser Applications, 2013. 25(5): 052006.

[903]   Bontha, S., et al., Thermal process maps for predicting solidification microstructure in laser fabrication of thin-wall structures. Journal of Materials Processing Technology, 2006. 178(1): 135–142.

[904]   Marshall, G.J., et al., Understanding the microstructure formation of Ti-6Al-4V during direct laser deposition via in-situ thermal monitoring. JOM, 2016. 68(3): 778–790.

[905]   Lewandowski, J.J. and M. Seifi, Metal additive manufacturing: a review of mechanical properties. Annual Review of Materials Research, 2016. 46(1): 151–186.

[906]   Aversa, R., et al., Processability of bulk metallic glasses. American Journal of Applied Sciences, 2017. 14.

[907]   Su, S. and Y. Lu, Laser directed energy deposition of Zr-based bulk metallic glass composite with tensile strength. Materials Letters, 2019. 247: 79–81.

[908]   Harun, W.S.W., et al., A review of powder additive manufacturing processes for metallic biomaterials. Powder Technology, 2018. 327: 128–151.

[909]  Shin, Y.C., et al., Predictive modeling capabilities from incident powder and laser to mechanical properties for laser directed energy deposition. Computational Mechanics, 2018. 61(5): 617–636.

[910]  Qian, M. and F.H. Froes, Titanium Powder Metallurgy: Science, Technology and Applications. 2015, Elsevier Science.

[911]  Heralić, A., A.-K. Christiansson and B. Lennartson, Height control of laser metal-wire deposition based on iterative learning control and 3D scanning. Optics and Lasers in Engineering, 2012. 50(9): 1230–1241.

[912]  Meng, L., et al., Wire electrochemical micromachining of metallic glass using a carbon nanotube fiber electrode. Journal of Alloys and Compounds, 2017. 709: 760–771.

[913]  Wang, C., et al., Wire based plasma arc and laser hybrid additive manufacture of Ti-6Al-4V. Journal of Materials Processing Technology, 2021. 293: 117080.

[914]  Cao, L., et al., Effect of laser energy density on defects behavior of direct laser depositing 24CrNiMo alloy steel. Optics & Laser Technology, 2019. 111: 541–553.

[915]  Yan, J., et al. Optimization of process parameters in laser engineered Net shaping (LENS) deposition of multi-materials. in ASME 2015 International Design Engineering Technical Conferences and Computers and Information in Engineering Conference. 2015. American Society of Mechanical Engineers Digital Collection.

[916]  Srinivas, V., U. Savitha and G. Jagan Reddy, Processing and Characterization of NiCr-YSZ Compositionally Graded Coatings on Superalloy using Laser Engineered Net Shaping (LENS). Materials Today: Proceedings, 2018. 5(13, Part 3): 27277–27284.

[917]  Atwood, C., et al. Laser engineered net shaping (LENS): a tool for direct fabrication of metal parts. in Proceedings of ICALEO. 1998.

[918]  Griffith, M., et al. Free form fabrication of metallic components using laser engineered net shaping (LENS). in Proceedings of the Solid Freeform Fabrication Symposium. 1996. University of Texas at Austin Austin, TX.

[919]  Zhu, Z. and S. Zhang, Large-scale fabrication of micro-lens array by novel end-fly-cutting-servo diamond machining. Opt Express, 2015. 23.

[920]  Jeantette, F.P., et al., Method and System for Producing Complex-Shape Objects. 2000, Google Patents.

[921]  Tejaswini, N., K.R. Babu and K.S. Ram. Functionally Graded Material: An Overview, in International Conference on Advances in Engineering Science and Management. 2015.

[922]  Brandt, M., Laser Additive Manufacturing: Materials, Design, Technologies, and Applications. 2016, Elsevier Science.

[923]  Tauqir, A., Transport Phenomena and Microstructural Developments During Electron Beam Melting (Rapid Solidification, Fluid Flow, Carbides, Tool Steel). 1986,

[924]  Sigl, M., S. Lutzmann and M. Zäh. Transient physical effects in electron beam sintering. in Solid Freeform Fabrication Symposium Proceedings. Austin, TX. 2006.

[925]  Kahnert, M., S. Lutzmann and M. Zaeh. Layer formations in electron beam sintering. in Solid freeform fabrication symposium. 2007.

[926]  Gong, X., T. Anderson and K. Chou. Review on powder-based electron beam additive manufacturing technology. in ASME/ISCIE 2012 international symposium on flexible automation. 2012. American Society of Mechanical Engineers.

[927]  Murr, L.E., Metallurgy of additive manufacturing: Examples from electron beam melting. Additive Manufacturing, 2015. 5: 40–53.

[928]  Vutova, K. and V. Donchev, Electron beam melting and refining of metals: computational modeling and optimization. Materials, 2013. 6(10): 4626.

[929]  Bergmann, H.-W. and B.L. Mordike, Laser and electron-beam melted amorphous layers. Journal of Materials Science, 1981. 16(4): 863–869.

[930]  Bhanumurthy, K., G.K. Dey and S. Banerjee, Metallic glass formation by solid state reaction in bulk zirconium – copper diffusion couples. Scripta Metallurgica, 1988. 22(9): 1395–1398.

[931]  Koptyug, A., L.-E. Rännar and M. Bäckström. Bulk Metallic Glass Manufacturing Using Electron Beam Melting. in International Conference on Additive Manufacturing & 3D Printing, Nottingham, UK, July 2013. 2013.

[932]  Drescher, P. and H. Seitz, Processability of an amorphous metal alloy powder by electron beam melting. RTeJournal – Fachforum für Rapid Technologie, 2015. 2015(1).

[933]  Sun, S.-H., et al., Build direction dependence of microstructure and high-temperature tensile property of Co–Cr–Mo alloy fabricated by electron beam melting. Acta Materialia, 2014. 64: 154–168.

[934]  Shifeng, W., et al., Effect of molten pool boundaries on the mechanical properties of selective laser melting parts. Journal of Materials Processing Technology, 2014. 214(11): 2660–2667.

[935]  Gusarov, A.V. and I. Smurov, Modeling the interaction of laser radiation with powder bed at selective laser melting. Physics Procedia, 2010. 5: 381–394.

[936]  Rafique, M.M.A., D. Qiu and M. Easton, Modeling and simulation of microstructural evolution in Zr based Bulk Metallic Glass Matrix Composites during solidification. MRS Advances, 2017. 2(58): 3591–3606.

[937]  Balla, V.K. and A. Bandyopadhyay, Laser processing of Fe-based bulk amorphous alloy. Surface and Coatings Technology, 2010. 205(7): 2661–2667.

[938]  Basu, A., et al., Laser surface coating of Fe-Cr-Mo-Y-B-C bulk metallic glass composition on AISI 4140 steel. Surface and Coatings Technology, 2008. 202(12): 2623–2631.

[939]  Yan, M., et al., The influence of topological structure on bulk glass formation in Al-based metallic glasses. Scripta Materialia, 2011. 65(9): 755–758.

[940]  Mu, J., et al., Synthesis and Properties of Al-Ni-La Bulk Metallic Glass. Advanced Engineering Materials, 2009. 11(7): 530–532.

[941]  Yang, B.J., et al., Al-rich bulk metallic glasses with plasticity and ultrahigh specific strength. Scripta Materialia, 2009. 61(4): 423–426.

[942]  Yang, B.J., et al., Developing aluminum-based bulk metallic glasses. Philosophical Magazine, 2010. 90(23): 3215–3231.

[943]  Balla, V.K., et al., Laser-assisted Zr/ZrO2 coating on Ti for load-bearing implants. Acta Biomaterialia, 2009. 5(7): 2800–2809.

[944]  Balla, V.K., et al., Direct laser processing of a tantalum coating on titanium for bone replacement structures. Acta Biomaterialia, 2010. 6(6): 2329–2334.

[945]  Wang, X., et al., Topological design and additive manufacturing of porous metals for bone scaffolds and orthopaedic implants: A review. Biomaterials, 2016. 83: 127–141.

[946]  Rafique, M.M.A., Modeling and Simulation of Heat Transfer Phenomena. 2015,

[947]  Maruyama, S., Molecular dynamics method for microscale heat transfer. Advances in numerical heat transfer, 2000. 2(6): 189–226.

[948]  Maruyama, S., A molecular dynamics simulation of heat conduction of a finite length single-walled carbon nanotube. Microscale Thermophysical Engineering, 2003. 7(1): 41–50.

[949]  Wang, X. and X. Xu, Molecular dynamics simulation of heat transfer and phase change during laser material interaction. Journal of Heat Transfer, 2001. 124(2): 265–274.

[950]  Krivtsov, A.M. and M. Wiercigroch, Molecular dynamics simulation of mechanical properties for polycrystal materials. Materials Physics and Mechanics, 2001. 3: 45–51.

[951]  Hao, T., et al., Molecular simulations of crystallization behaviors of polymers grafted on two-dimensional filler. Polymer, 2016. 100: 10–18.

[952]  Broecker, P. and S. Trebst, Entanglement and the fermion sign problem in auxiliary field quantum Monte Carlo simulations. Physical Review B, 2016. 94(7): 075144.

[953]  Howell, J.R., The Monte Carlo method in radiative heat transfer. Journal of Heat Transfer, 1998. 120(3): 547–560.

[954]  Zeeb, C.N., Performance and Accuracy Enhancements of Radiative Heat Transfer Modeling via Monte Carlo. 2002, Citeseer.

[955]  Planas Almazan, P. Accuracy of Monte Carlo ray-tracing thermal radiation calculations: A practical discussion, in Sixth European Symposium on Space Environmental Control Systems. 1997.

[956]  Modest, M.F., Backward Monte Carlo simulations in radiative heat transfer. Journal of Heat Transfer, 2003. 125(1): 57–62.

[957]  Frijns, A.J.H., et al., Molecular Dynamics and Monte Carlo Simulations for Heat Transfer in Micro and Nano-channels, in Computational Science – ICCS 2004: 4th International Conference, Kraków, Poland, June 6–9, 2004, Proceedings, Part IV, M. Bubak, et al., Editors. 2004, Springer Berlin Heidelberg: Berlin, Heidelberg. p. 661–666.

[958]  Westhoff, D., et al., Stochastic modeling and predictive simulations for the microstructure of organic semiconductor films processed with different spin coating velocities. Modelling and Simulation in Materials Science and Engineering, 2015. 23(4): 045003.

[959]  Masrour, R. and A. Jabar, Magnetic properties of dendrimer structures with different coordination numbers: A Monte Carlo study. Journal of Magnetism and Magnetic Materials, 2016. 417: 397–400.

[960]  Binder, K. and P. Virnau, Overview: understanding nucleation phenomena from simulations of lattice gas models. The Journal of Chemical Physics, 2016. 145(21): 211701.

[961]  Guo, G.-Q., et al., Structure-induced microalloying effect in multicomponent alloys. Materials & Design, 2016. 103: 308–314.

[962]  Nandi, U.K., et al., Composition dependence of the glass forming ability in binary mixtures: The role of demixing entropy. The Journal of Chemical Physics, 2016. 145(3): 034503.

[963]  Antonie, O., Langevin Dynamics Of Protein Folding: Influence of Confinement. 2007, Masters De-gree Project, Scuola Internazionale Superiore di Studi Avanzati, SISSA: Trieste, Italy.

[964]  Snook, I., The Langevin and Generalised Langevin approach to the Dynamics of Atomic, Polymeric and Colloidal Systems. 2006, Elsevier.

[965]  Busetti, F., Simulated Annealing Overview. 2003, JP Morgan: Italy.

[966]  Front Matter A2 – CHRISTIAN, J.W, in The Theory of Transformations in Metals and Alloys. 2002, Pergamon: Oxford. p. iii.

[967]  Gránásy, L., Quantitative analysis of the classical nucleation theory on glass-forming alloys. Journal of Non-Crystalline Solids, 1993. 156: 514–518.

[968]  Ablitzer, D., Transport phenomena and modelling in melting and refining processes. Journal of Physics, 1993. 03(C7): C7-873-C7-882.

[969]  Gandin, C.-A., et al., A three-dimensional cellular automation-finite element model for the prediction of solidification grain structures. Metallurgical and Materials Transactions A, 1999. 30(12): 3153–3165.

[970]  Guillemot, G., et al., A new cellular automaton – finite element coupling scheme for alloy solidification. Modelling and Simulation in Materials Science and Engineering, 2004. 12(3): 545.

[971]  Hebi, Y. and D.F. Sergio, A cellular automaton model for dendrite growth in magnesium alloy AZ91. Modelling and Simulation in Materials Science and Engineering, 2009. 17(7): 075011.

[972]  Kremeyer, K., Cellular Automata Investigations of Binary Solidification. Journal of Computational Physics, 1998. 142(1): 243–263.

[973]  Tian, F., Z. Li and J. Song, Solidification of laser deposition shaping for TC4 alloy based on cellular automation. Journal of Alloys and Compounds, 2016. 676: 542–550.

[974]  Zhang, X., et al., A three-dimensional cellular automaton model for dendritic growth in multi-component alloys. Acta Materialia, 2012. 60(5): 2249–2257.

[975]  Zaeem, M.A., H. Yin and S.D. Felicelli, Comparison of Cellular Automaton and Phase Field Models to Simulate Dendrite Growth in Hexagonal Crystals. Journal of Materials Science & Technology, 2012. 28(2): 137–146.

[976]  Krone, S.M., Spatial models: stochastic and deterministic. Mathematical and Computer Modelling, 2004. 40(3): 393–409.

[977]  Bao, Y.B. and J. Meskas, Lattice Boltzmann Method for Fluid Simulations. 2011, Department of Mathematics, Courant Institute of Mathematical Sciences, New York University.

[978]  Raabe, D., Overview of the lattice Boltzmann method for nano- and microscale fluid dynamics in materials science and engineering. Modelling and Simulation in Materials Science and Engineering, 2004. 12(6): R13.

[979]  Sun, D., et al., Lattice Boltzmann modeling of dendritic growth in a forced melt convection. Acta Materialia, 2009. 57(6): 1755–1767.

[980]  Sun, D.K., et al., Modelling of dendritic growth in ternary alloy solidification with melt convection. International Journal of Cast Metals Research, 2011. 24(3–4): 177–183.

[981]  Sun, D.K., et al., Lattice Boltzmann modeling of dendritic growth in forced and natural convection. Computers & Mathematics with Applications, 2011. 61(12): 3585–3592.

[982]  Eshraghi, M., B. Jelinek and S.D. Felicelli, Large-scale three-dimensional simulation of dendritic solidification using lattice Boltzmann method. JOM, 2015. 67(8): 1786–1792.

[983]  Asle Zaeem, M., Advances in modeling of solidification microstructures. JOM, 2015. 67(8): 1774–1775.

[984]  Böttger, B., J. Eiken and I. Steinbach, Phase field simulation of equiaxed solidification in technical alloys. Acta Materialia, 2006. 54(10): 2697–2704.

[985]  Emmerich, H., Phase-field models for colloidal systems: First steps, perspectives and open challenges. Current Opinion in Solid State and Materials Science, 2011. 15(3): 83–86.

[986]  Zhang, R., et al., Phase-field simulation of solidification in multicomponent alloys coupled with thermodynamic and diffusion mobility databases. Acta Materialia, 2006. 54(8): 2235–2239.

[987]  Nestler, B. and A. Choudhury, Phase-field modeling of multi-component systems. Current Opinion in Solid State and Materials Science, 2011. 15(3): 93–105.

[988]  Choudhury, A., et al., Comparison of phase-field and cellular automaton models for dendritic solidification in Al–Cu alloy. Computational Materials Science, 2012. 55: 263–268.

[989]  Wang, W., S. Luo and M. Zhu, Numerical simulation of three-dimensional dendritic growth of alloy: Part II – model application to Fe-0.82WtPctC alloy. Metallurgical and Materials Transactions A, 2016. 47(3): 1355–1366.

[990]  Zhu, M.F. and D.M. Stefanescu, Virtual front tracking model for the quantitative modeling of dendritic growth in solidification of alloys. Acta Materialia, 2007. 55(5): 1741–1755.

[991]  Dijkhuizen, W., et al., DNS of gas bubbles behaviour using an improved 3D front tracking model – Model development. Chemical Engineering Science, 2010. 65(4): 1427–1437.

[992]  Chen, C., M. Miller and K. Sepehrnoori, A New 3D Front-Tracking Approach to Modeling Displacement Processes in Complex Large-Scale Reservoirs, in SPE Reservoir Simulation Symposium. 1997, Society of Petroleum Engineers.

[993]  Song, K.J., et al., Virtual front tracking cellular automaton modeling of isothermal β to α phase transformation with crystallography preferred orientation of TA15 alloy. Modelling and Simulation in Materials Science and Engineering, 2014. 22(1): 015006.

[994]   Wang, Y., et al., Sharp interface model for solid-state dewetting problems with weakly anisotropic surface energies. Physical Review B, 2015. 91(4): 045303.

[995]   Vermolen, F., et al. A suite of sharp interface models for solid-state transformations in alloys. in ECCOMAS CFD 2006: Proceedings of the European Conference on Computational Fluid Dynamics, Egmond aan Zee, The Netherlands, September 5–8, 2006. 2006. Delft University of Technology; European Community on Computational Methods in Applied Sciences (ECCOMAS).

[996]   Vermolen, F.J., Zener Solutions for Particle Growth in Multi-Component Alloys. 2006, Delft University of Technology, Faculty of Electrical Engineering, Mathematics and Computer Science, Delft Institute of Applied Mathematics.

[997]   Ågren, J., Numerical treatment of diffusional reactions in multicomponent alloys. Journal of Physics and Chemistry of Solids, 1982. 43(4): 385–391.

[998]   Caginalp, G. and W. Xie, Phase-field and sharp-interface alloy models. Physical Review E, 1993. 48(3): 1897–1909.

[999]   Nielsen, C.P. and H. Bruus, Sharp-interface model of electrodeposition and ramified growth. Physical Review E, 2015. 92(4): 042302.

[1000]  Sato, Y. and B. Ničeno, A sharp-interface phase change model for a mass-conservative interface tracking method. Journal of Computational Physics, 2013. 249: 127–161.

[1001]  Francois, M.M., et al., A balanced-force algorithm for continuous and sharp interfacial surface tension models within a volume tracking framework. Journal of Computational Physics, 2006. 213(1): 141–173.

[1002]  Tang, C. and C.H. Wong, Effect of atomic-level stresses on local dynamic and mechanical properties in CuxZr100−x metallic glasses: A molecular dynamics study. Intermetallics, 2015. 58: 50–55.

[1003]  Feng, S.D., et al., Transition from elasticity to plasticity in Zr35Cu65 metallic glasses: A molecular dynamics study. Journal of Non-Crystalline Solids, 2015. 430: 94–98.

[1004]  Falk, M.L., Molecular-dynamics study of ductile and brittle fracture in model noncrystalline solids. Physical Review B, 1999. 60(10): 7062–7070.

[1005]  Metropolis, N., et al., Equation of state calculations by fast computing machines. The Journal of Chemical Physics, 1953. 21(6): 1087–1092.

[1006]  Wood, W.W. and J.D. Jacobson, Preliminary results from a recalculation of the Monte Carlo equation of state of hard spheres. The Journal of Chemical Physics, 1957. 27(5): 1207–1208.

[1007]  Kroese, D.P., et al., Why the Monte Carlo method is so important today. Wiley Interdisciplinary Reviews: Computational Statistics, 2014. 6(6): 386–392.

[1008]  Pusztai, L. and E. Sváb, Structure study of Ni62Nb38 metallic glass using reverse Monte Carlo simulation. Journal of Non-Crystalline Solids, 1993. 156: 973–977.

[1009]  Hu, Y., et al., Monte Carlo simulation of dual magnetic phase behavior in bulk metallic glasses. Journal of Magnetism and Magnetic Materials, 2010. 322(17): 2567–2570.

[1010]  Yang, M.H., J.H. Li and B.X. Liu, Proposed correlation of structure network inherited from producing techniques and deformation behavior for Ni-Ti-Mo metallic glasses via atomistic simulations. Scientific Reports, 2016. 6: 29722.

[1011]  Huang, L., et al., Surface engineering of a Zr-based bulk metallic glass with low energy Ar- or Ca-ion implantation. Materials Science and Engineering: C, 2015. 47: 248–255.

[1012]  Gilbert, A., Introduction to Computational Quantum Chemistry: Theory. 2007, University Lecture.

[1013]  Parr, R.G., D.P. Craig and I.G. Ross, Molecular orbital calculations of the lower excited electronic levels of benzene, configuration interaction included. The Journal of Chemical Physics, 1950. 18(12): 1561–1563.

[1014] Sha, Z.D. and Q.X. Pei, Ab initio study on the electronic origin of glass-forming ability in the binary Cu–Zr and the ternary Cu–Zr–Al(Ag) metallic glasses. Journal of Alloys and Compounds, 2015. 619: 16–19.

[1015] Kirklin, S., The Open Quantum Materials Database (OQMD): assessing the accuracy of DFT formation energies. npj Computational Materials, 2015. 1.

[1016] Sarhaddi, R., H. Arabi and F. Pourarian, Structural, stability and electronic properties of C15-AB2 (A = Ti Zr; B = Cr) intermetallic compounds and their hydrides: An ab initio study. International Journal of Modern Physics B, 2014. 28(17): 1450105.

[1017] Sun, S.-J., et al., Effects of Strontium incorporation to Mg-Zn-Ca biodegradable bulk metallic glass investigated by molecular dynamics simulation and density functional theory calculation. Scientific Reports, 2020. 10(1): 2515.

[1018] Zheng, G.-P., A density functional theory study on the deformation behaviors of Fe-Si-B metallic glasses. International Journal of Molecular Sciences, 2012. 13(8).

[1019] Wang, J., et al., Compressibility and hardness of Co-based bulk metallic glass: A combined experimental and density functional theory study. Applied Physics Letters, 2011. 99(15): 151911.

[1020] Ju, S.-P. and -C.-C. Yang, Understanding the structural, mechanical, thermal, and electronic properties of MgCa bulk metallic glasses by molecular dynamics simulation and density functional theory calculation. Computational Materials Science, 2018. 154: 256–265.

[1021] Daw, M.S. and M.I. Baskes, Embedded-atom method: Derivation and application to impurities, surfaces, and other defects in metals. Physical Review B, 1984. 29(12): 6443.

[1022] Foiles, S., M. Baskes and M.S. Daw, Embedded-atom-method functions for the fcc metals Cu, Ag, Au, Ni, Pd, Pt, and their alloys. Physical Review B, 1986. 33(12): 7983.

[1023] Baskes, M.I., Modified embedded-atom potentials for cubic materials and impurities. Physical Review B, 1992. 46(5): 2727–2742.

[1024] Ercolessi, F., M. Parrinello and E. Tosatti, Simulation of gold in the glue model. Philosophical Magazine A, 1988. 58(1): 213–226.

[1025] Donald, W.B., et al., A second-generation reactive empirical bond order (REBO) potential energy expression for hydrocarbons. Journal of Physics: Condensed Matter, 2002. 14(4): 783.

[1026] Pronk, S., et al., GROMACS 4.5: a high-throughput and highly parallel open source molecular simulation toolkit. Bioinformatics, 2013.

[1027] Seabaugh, A., The Tunneling Transistor. IEEE Spectrum, 2013. 50(10): 35–62.

[1028] Schulz, R., et al., Scaling of multimillion-atom biological molecular dynamics simulation on a petascale supercomputer. Journal of Chemical Theory and Computation, 2009. 5(10): 2798–2808.

[1029] Zheng, B., et al., Thermal Behavior and Microstructural Evolution during Laser Deposition with Laser-Engineered Net Shaping: Part I. Numerical Calculations. Metallurgical and Materials Transactions A, 2008. 39(9): 2228–2236.

[1030] Wang, Y., et al., Research on modeling of heat source for electron beam welding fusion-solidification zone. Chinese Journal of Aeronautics, 2013. 26(1): 217–223.

[1031] Kik, T., Heat source models in numerical simulations of laser welding. Materials, 2020. 13(11).

[1032] Al Hamahmy, M.I. and I. Deiab, Review and analysis of heat source models for additive manufacturing. The International Journal of Advanced Manufacturing Technology, 2020. 106(3): 1223–1238.

[1033] Rosenthal, D., The theory of moving sources of heat and its application of metal treatments. Transactions of ASME, 1946. 68: 849–866.

[1034] Rosenthal, D., Mathematical theory of heat distribution during welding and cutting. Welding Journal, 1941. 20: 220–234.

[1035] Pavelic, V., Experimental and computed temperature histories in gas tungsten arc welding of thin plates. Welding Journal Research Supplement, 1969. 48: 296–305.

[1036] Tekriwal, P. and J. Mazumder, Finite element analysis of three-dimensional transient heat transfer in GMA welding. Welding Journal, 1988. 67(7): 150s–156s.

[1037] Friedman, E., Thermomechanical analysis of the Welding Process using the Finite Element Method. 1975,

[1038] Paley, Z. and P. Hibbert, Computation of temperatures in actual weld designs. Welding journal, 1975. 54(11): 385s–392s.

[1039] Goldak, J., A. Chakravarti and M. Bibby, A new finite element model for welding heat sources. Metallurgical Transactions B, 1984. 15(2): 299–305.

[1040] Dezfoli, A.R.A., et al., Determination and controlling of grain structure of metals after laser incidence: Theoretical approach. Scientific Reports, 2017. 7(1): 41527.

[1041] Rafique, M.M.A. and J. Iqbal, Modeling and simulation of heat transfer phenomena during investment casting. International Journal of Heat and Mass Transfer, 2009. 52(7–8): 2132–2139.

[1042] Bennon, W.D. and F.P. Incropera, A continuum model for momentum, heat and species transport in binary solid-liquid phase change systems – I. Model formulation. International Journal of Heat and Mass Transfer, 1987. 30(10): 2161–2170.

[1043] Voller, V.R., A.D. Brent and C. Prakash, The modelling of heat, mass and solute transport in solidification systems. International Journal of Heat and Mass Transfer, 1989. 32(9): 1719–1731.

[1044] Ganesan, S. and D.R. Poirier, Conservation of mass and momentum for the flow of interdendritic liquid during solidification. Metallurgical Transactions B, 1990. 21(1): 173–181.

[1045] Ni, J. and C. Beckermann, A volume-averaged two-phase model for transport phenomena during solidification. Metallurgical Transactions B, 1991. 22(3): 349–361.

[1046] Rappaz, M., Modelling of microstructure formation in solidification processes. International Materials Reviews, 1989. 34(1): 93–124.

[1047] Kurz, W., B. Giovanola and R. Trivedi, Theory of microstructural development during rapid solidification. Acta Metallurgica, 1986. 34(5): 823–830.

[1048] Trivedi, R., P. Magnin and W. Kurz, Theory of eutectic growth under rapid solidification conditions. Acta Metallurgica, 1987. 35(4): 971–980.

[1049] Basak, A., R. Acharya and S. Das, Additive manufacturing of single-crystal superalloy CMSX-4 through scanning laser epitaxy: computational modeling, experimental process development, and process parameter optimization. Metallurgical and Materials Transactions A, 2016. 47(8): 3845–3859.

[1050] Christian, J.W., CHAPTER 10 – The Classical Theory of Nucleation, in The Theory of Transformations in Metals and Alloys. 2002, Pergamon: Oxford. 422–479.

[1051] Avrami, M., Kinetics of phase change. I general theory. The Journal of Chemical Physics, 1939. 7(12): 1103–1112.

[1052] Price, C.W., Simulations of grain impingement and recrystallization kinetics. Acta Metallurgica, 1987. 35(6): 1377–1390.

[1053] Zou, J., Simulation de la solidification eutectique équiaxe. 1989,

[1054] Thévoz, P., J.L. Desbiolles and M. Rappaz, Modeling of equiaxed microstructure formation in casting. Metallurgical Transactions A, 1989. 20(2): 311–322.

[1055] Stefanescu, D., Science and Engineering of Casting Solidification. 2015, Springer.

[1056] Kurz, W. and D.J. Fisher, Fundamentals of Solidification. 1986, Trans Tech Publications.

[1057] Chalmers, B. Principles of Solidification, in Applied Solid State Physics, Low, W. and M. Schieber, Editors. 1970, Springer US: Boston, MA. 161–170.

[1058]  Rappaz, M. and E. Blank, Simulation of oriented dendritic microstructures using the concept of dendritic lattice. Journal of Crystal Growth, 1986. 74(1): 67–76.

[1059]  Spittle, J.A. and S.G.R. Brown, Computer simulation of the effects of alloy variables on the grain structures of castings. Acta Metallurgica, 1989. 37(7): 1803–1810.

[1060]  Brown, S.G.R. and J.A. Spittle, Computer simulation of grain growth and macrostructure development during solidification. Materials Science and Technology, 1989. 5(4): 362–368.

[1061]  Anderson, M.P., et al., Computer simulation of grain growth – I. Kinetics. Acta Metallurgica, 1984. 32(5): 783–791.

[1062]  González, S., Role of minor additions on metallic glasses and composites. Journal of Materials Research, 2015. 31(1): 76–87.

[1063]  Song, H., et al., Simulation Study of Heterogeneous Nucleation at Grain Boundaries During the Austenite-Ferrite Phase Transformation: Comparing the Classical Model with the Multi-Phase Field Nudged Elastic Band Method, In: Metallurgical and Materials Transactions A. 2016. 1–9.

[1064]  Hunt, J.D., Steady state columnar and equiaxed growth of dendrites and eutectic. Materials Science and Engineering, 1984. 65(1): 75–83.

[1065]  Tan, W., N.S. Bailey and Y.C. Shin, A novel integrated model combining Cellular Automata and Phase Field methods for microstructure evolution during solidification of multi-component and multi-phase alloys. Computational Materials Science, 2011. 50(9): 2573–2585.

[1066]  StJohn, D.H., et al., The Interdependence Theory: The relationship between grain formation and nucleant selection. Acta Materialia, 2011. 59(12): 4907–4921.

[1067]  Maxwell, I. and A. Hellawell, A simple model for grain refinement during solidification. Acta Metallurgica, 1975. 23(2): 229–237.

[1068]  Acharya, R., et al. Computational Modeling and Experimental Validation of Microstructural Development in Superalloy CMSX-4 Processed Through Scanning Laser Epitaxy, in Solid Freeform Fabrication Symposium. 2012.

[1069]  Blázquez, J.S., et al., Instantaneous growth approximation describing the nanocrystallization process of amorphous alloys: A cellular automata model. Journal of Non-Crystalline Solids, 2008. 354(30): 3597–3605.

[1070]  Wang, W., S. Luo and M. Zhu, Numerical Simulation of Three-Dimensional Dendritic Growth of Alloy: Part I – Model Development and Test. Metallurgical and Materials Transactions A, 2016. 47(3): 1339–1354.

[1071]  Perim, E., et al., Spectral descriptors for bulk metallic glasses based on the thermodynamics of competing crystalline phases. Nature Communications, 2016. 7: 12315.

[1072]  Wang, X.D., et al., Atomic picture of elastic deformation in a metallic glass. Scientific Reports, 2015. 5: 9184.

[1073]  Zheng, G.-P., A Density functional theory study on the deformation behaviors of Fe-Si-B metallic glasses. International Journal of Molecular Sciences, 2012. 13(8): 10401–10409.

[1074]  Inoue, A., Bulk glassy alloys: historical development and current research. Engineering, 2015. 1(2): 185–191.

[1075]  Inoue, A., T. Zhang and E. Makabe, Production Methods of metallic Glasses by a Suction Casting Method. 1998, Google Patents.

[1076]  Greer, A.L., Liquid metals: Supercool order. Nature Materials, 2006. 5(1): 13–14.

[1077]  Liu, X.J., et al., Metallic liquids and glasses: atomic order and global packing. Physical Review Letters, 2010. 105(15): 155501.

[1078]  Wang, W.H., Metallic glasses: Family traits. Nature Materials, 2012. 11(4): 275–276.

[1079]  Kumar, G., A. Desai and J. Schroers, Bulk metallic glass: the smaller the better. Advanced Materials, 2011. 23(4): 461–476.

[1080]  Wang, C.Y. and C. Beckermann, A multiphase solute diffusion model for dendritic alloy solidification. Metallurgical Transactions A, 1993. 24(12): 2787–2802.

[1081]  Li, H.Q., J.H. Yan and H.J. Wu, Modelling and simulation of bulk metallic glass production process with suction casting. Materials Science and Technology, 2009. 25(3): 425–431.

[1082]  Lindgren, L.-E., et al., Simulation of additive manufacturing using coupled constitutive and microstructure models. Additive Manufacturing.

[1083]  Markl, M. and C. Körner, Multiscale modeling of powder bed–based additive manufacturing. Annual Review of Materials Research, 2016. 46(1): 93–123.

[1084]  Fan, Z., et al. Numerical simulation of the evolution of solidification microstructure in laser deposition, in Proceedings of the 18th Annual Solid Freeform Fabrication Symposium. 2007.

[1085]  Zhu, M.F., S.Y. Lee and C.P. Hong, Modified cellular automaton model for the prediction of dendritic growth with melt convection. Physical Review E, 2004. 69(6): 061610.

# Exercise

## Subjective questions

1. Define metallic glass and metallic glass matrix composite and give examples.
2. Describe classification of bulk metallic glasses and give examples. Describe the important characteristics of bulk metallic glasses. Explain how they differentiate them from other materials.
3. Describe various methods of production of bulk metallic glasses. Highlight each salient features, uniqueness, advantages and disadvantages.
4. What is glass transition temperature, crystallization temperature and liquid temperature? Differentiate. Give examples.
5. Define vitrification, devitrification, relaxation, rejuvenation, phase separation and liquid–liquid transition. Explain them with reference to bulk metallic glasses and their types. Support your answer with examples.
6. Describe which factors are mostly responsible for evolution of crystalline microstructure in glass when making metallic glass matrix composites? Explain your answer and give examples. [Hint: Only consider liquid-to-solid transformation.]
7. What is shear band? How can it be observed? What is its role in deformation in metallic glasses? How it is different from dislocations and how can it play a part in increasing toughness of bulk metallic glasses? Give examples and illustrations.
8. Name few methods to characterize metallic glasses. Which one you would choose if you have to observe percentage crystallinity in otherwise 100% glass? Give examples and justify your response.

## Quantitative questions and numerical

9. Suppose you are going to cast a wedge shaped sample of $Zr_{65}Cu_{15}Al_{10}Ni_{10}$ with the following dimensions:
   i.   Length = 2 in
   ii.  Width at top = 0.25 in
   iii. Thickness = 0.25 in

   Select a suitable die material for casting this. Calculate charge and propose the final weight of casting to be melted accounting material and machine losses. Propose the safety factor.

10. Suppose you are given an optical micrograph of metallic glass matrix composite (Figure 39). How would you determine it is a metallic glass or composite? Select a suitable method to calculate crystallinity. Calculate the volume fraction of crystal phases using this method.

https://doi.org/10.1515/9783110747232-013

11. For a laser of fixed spot size and energy striking the surface of bulk metallic glass ($Zr_{47.5}Cu_{45.5}Al_5Co_2$), derive an expression for time it takes to complete one roster.

12. For question 11, suppose spot size = 0.4 mm and energy = 500 W, calculate the scan speed of laser traversing the sample surface.

13. For question 11, suppose you are going to model this process. Write moving heat source equation you will choose and describe which program you will choose. Justify your answer with argument and reasoning.

14. For question 11, write an equation of thermal conductivity for conduction phenomena (assume steady-state conditions). Write equation for thermal diffusivity and explain the difference between two and their significance.

15. For question 14, suppose plate thickness = 10 mm, plate shape is square with one side = 8 mm, calculate heat flux traveling across the plate from 90 °C to room temperature (assume the missing value for Zr).

16. Suppose you are looking to determine the heat transfer pattern of laser striking the sample surface at a specific point. Out of below, which method you would choose and why?
    i. MATLAB,   ii. FORTRAN,   iii. Ansys   iv. ABAQUS
    v. ABAQUS with subroutines

17. Suppose you choose ABAQUS with subroutines. Describe which subroutine you will choose for the above problem. Please write the script of your subroutine.

# Design problems

1. Glass-forming ability.
   a. What is glass-forming ability? Explain its significance and give examples of systems highlighting its importance.
   b. Name three glass-forming systems with excellent glass-forming ability.
   c. Describe methods to determine glass-forming ability and give mathematical expression for each.
2. Phase diagrams
   a. What are phase diagrams? What are their types and classification? What are common phase diagrams for bulk metallic glasses?
   b. Why phase diagrams are called equilibrium phase diagrams? How they differ from nonequilibrium (time–temperature transformation, continuous cooling transformation and isothermal cooling transformation [ICT]) phase diagrams?
   c. For a binary phase diagram, what are phase rule and lever rule? What is their importance and how they play a part in understanding phase formation and transformation? Give examples.
   d. What are equilibrium, free energy, enthalpy, heat capacity and entropy? Explain these for a binary phase diagram of Zr–Cu metallic glass.
   e. What are critical temperatures and critical temperature lines for a binary phase diagram and what is the critical temperature surface for a ternary phase diagram? Draw and give examples for Zr–Cu, Zr–Cu–Al and Zr–Cu–Ni systems.
   f. For binary and ternary phase diagrams, define and explain invariant points and reactions. For a ternary phase diagram, draw and explain the vertical section. Explain it with an example from industry.
   g. For a ternary phase diagram, define and explain space model, isothermal sections and projected view.
3. Design of Mars rover wheel and axle
   a. Suppose you are designing next Mars rover. Derive an expression for
      i. Tensile strength of a material for stresses
         1. Normal to a plane perpendicular to the direction of rotation at point P ($\sigma_n$) (Figure 43)
         2. Parallel to a plane perpendicular to the direction of rotation (V) at point P ($\sigma_p$) (Figure 43)
         3. At an angle $\theta$ to a plane perpendicular to the direction of rotation at point ($\sigma_o$) P (Figure 43) [hint: use trigonometric identities for horizontal and normal coordinates of a vector]

https://doi.org/10.1515/9783110747232-014

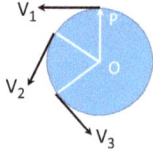

**Figure 43:** Schematic rover wheel.

where OP is the distance from origin (O) to a point P at the surface of a wheel.

ii. Compressive stresses at the surface (skin stresses) and in bulk

iii. Torsional strength of a material of an axle joining two wheels

iv. Based on these stresses and envelope conditions of planet, select the

    1. material of construction

    2. method of manufacturing and fabrication (if any)

    3. postprocessing, reasoning and justification

[Hint] Conditions at Mars: $g = 1/3\ g_e$, composition: $CO_2 > 96\%$, $N_2 = 2\%$, $Ar = 2\%$, temperature = cryogenic (minimum: $-140\ °C$, max: $30\ °C$, average: $-63\ °C$).

# Index

https://doi.org/10.1515/9783110747232-015